Population, Land Management, and Environmental Change

The UNU Global Environmental Forum IV, Population, Land Management, and Environmental Change, 25 May 1995, was organized by the United Nations University in cooperation with the United Nations Environment Programme. It was supported by the Global Environment Centre Foundation and sponsored by Obayashi Corporation.

Population, Land Management, and Environmental Change

UNU Global Environmental Forum IV

Edited by Juha I. Uitto and Akiko Ono

The United Nations University
Tokyo, Japan

The views expressed in this publication are those of the authors and do not necessarily reflect the views of the United Nations University.

The United Nations University, 53-70, Jingumae 5-chome,
Shibuya-ku, Tokyo 150, Japan
Tel: (03) 3499-2811   Fax: (03) 3406-7345
Telex: J25442   Cable: UNATUNIV TOKYO

UNU Office in North America
2 United Nations Plaza, Room DC2-1462-70, New York, NY 10017
Tel: (212) 963-6387   Fax: (212) 371-9454   Telex: 422311 UN UI

Typeset by Asco Trade Typesetting Limited, Hong Kong
Printed by Permanent Typesetting and Printing Co., Ltd., Hong Kong
Cover design by Takashi Suzuki

UNUP-956
ISBN 92-808-0956-3
01500 P

Contents

Preface

Juha I. Uitto

The United Nations University is an autonomous academic organization under the United Nations umbrella, with headquarters in Tokyo. According to its charter, the University's mandate is to carry out research, postgraduate education, and dissemination of knowledge related to the pressing global problems facing mankind. It has also a specific mandate towards capacity building in the developing countries to assist them in dealing with the questions of development. An international community of scholars, the United Nations University works through networks of scientists from each of the world's continents. These networks serve to focus the issues to be researched and to bring together the brightest minds to cooperate in finding solutions to the international problems.

The purpose of the UNU Global Environmental Forum series is to disseminate research results on issues pertaining to global environmental change to a broader public. Since the beginning of the series in 1991, the Fora have covered a wide variety of issues. The first one, entitled "Monitoring and Action for the Earth," was concerned with new technologies for monitoring and observation of the changes occurring in the terrestrial and oceanic ecosystems. The second Forum, in 1993, focused on environmental change affecting rainforests and drylands in South-East Asia, Sub-Saharan Africa, and the Chinese drylands. The third Forum's title was posed in the form of a question: "Will Tropical Forests Change in a Global Greenhouse?" It examined the complex and multi-faceted interlinkages between global warming and the sustainability of tropical rainforest ecosystems and their biological diversity.

The fourth UNU Global Environmental Forum took place in Osaka, Japan, on 25 May 1995. Its theme was "Population, Land Management, and

Environmental Change." The Forum focused on the research carried out under the University's international collaborative research programme with the same title. This publication reproduces the papers presented at the Forum in an edited form. The first three papers set the objectives of the research programme (Harold Brookfield) and outline some of the underlying key concepts, including farmers' participation (Michael Stocking) and the role of women (Janet Momsen). The following four chapters highlight preliminary results from the research undertaken within the field research clusters of the programme in Papua New Guinea (Graham Sem and Ryutaro Ohtsuka), northern Thailand (Kanok Rerkasem), and the Amazon (Christine Padoch). The final chapter by Shunji Murai presents a different perspective to the population carrying capacity at the global level.

The Forum was organized jointly with the International Environmental Technology Centre of the United Nations Environment Programme (UNEP/IETC), and in cooperation with the Global Environmental Centre Foundation. The Forum, as well as all the earlier ones, was sponsored by Obayashi Corporation.

Opening Remarks

Takashi Inoguchi

It is my great pleasure to welcome you to the Global Environmental Forum organized by the United Nations University. Although this is already the fourth Forum, it is the first one to be held in the Kansai region. Therefore, it is of special importance to us to address the audience here today.

As many of you realize, this conference was originally scheduled to be held on 19 January 1995, but the Great Hanshin Earthquake that occurred just two days earlier on the other side of Osaka Bay so tragically forced us to postpone the event. We sincerely regret the change in plans and any inconvenience it may have caused you. However, obviously the postponement of the Forum was only an insignificant consequence of the earthquake disaster that brought so much destruction and suffering to the inhabitants in the Kobe area.

Environmental issues and the quest for sustainable development are at the centre of international attention today. The global environment is under increasing stress due to human activities. The world is experiencing an unprecedented growth in human population, associated with rapid economic growth, industrialization, and urbanization. Environmental problems can be broadly divided into two types: those phenomena that are truly global, affecting the earth as a whole, such as global warming or the depletion of the atmospheric ozone layer; and those with more localized, yet severe effects. The latter are typically expressed in deforestation, loss of topsoils due to erosion, and the conversion of agricultural land to urban land uses. There is urgent need to improve mankind's understanding of these complicated phenomena and their complex interlinkages. Scientific knowledge is required to back up the concrete actions needed at local, regional, and global levels to halt environmental degradation and move towards sustainable development.

One of the main foci of the United Nations University's research and training programme is environment and sustainable development. Following the United Nations Conference on Environment and Development (UNCED), or the Earth Summit, held in Rio de Janeiro three years ago, UNU appointed a high-level advisory team to prepare a plan of action for the University in response to the recommendations of the Earth Summit. The ensuing document, the UNU Agenda 21, defines the scope and focus of our environmental research and capacity-building activities in view of our specific strengths and comparative advantages, as well as the need for complimentarity and co-operation with other actors in the field.

Today's Forum is entitled "Population, Land Management, and Environmental Change," a title it borrows from a large international collaborative research programme carried out under UNU. The PLEC programme was initiated around two years ago to study the sustainability, or lack of it, of agricultural systems in smallholder areas in the tropical and subtropical parts of the world. These agricultural systems, essential for feeding an ever increasing population, are under severe pressure from multiple sources, including population growth, commercialization of the economy, and land-use changes. PLEC examines the impacts of land management practices on the sustainability of agricultural systems. It also studies the conservation of biological diversity in the managed agricultural ecosystems. The basic premise of the research is that many of the indigenous agricultural systems are highly adapted to the local ecosystems and are very conservation oriented. However, they are rapidly disappearing due to socio-economic changes in the society. The research programme works through a network of locally based research clusters around the world, covering a range of locations from the Brazilian Amazonia and the Caribbean, through West and East Africa, to South-East Asia. Key scientists involved in the PLEC programme are present here today and will give you more detailed information on the various aspects of research and the preliminary hypotheses and findings.

I would like to take this opportunity to acknowledge the role of our partners in organizing this Forum. UNEP, the United Nations Environment Programme, is our co-organizer. They have recently set up a new International Environmental Technology Centre here in Kansai, which will certainly play an important role in the transfer of environmental technologies and capacity building in developing countries. We also wish to thank the Global Environmental Centre Foundation for cooperating with us in making this Forum happen.

Last but not least, I wish to express our gratefulness to the sponsors of the UNU Global Environmental Forum series, Obayashi Corporation. Since 1991, Obayashi has supported UNU in the dissemination of research results on the important topics of global environmental change and sustainable development, and our cooperation has become an institutional feature in the UNU programme. It is our sincere hope that this cooperation will be carried on long into the future.

Welcoming Address

Richard A. Meganck

I feel greatly honoured to have been invited to address this distinguished gathering in Osaka on behalf of the United Nations Environment Programme.

Over the past 30 years we have observed in many parts of the world rapid population growth, mismanagement of land and water resources leading to low agricultural performance, and increasing environmental degradation. Are these disturbing trends connected? New studies confirm that they are. For example, Africa's population, agriculture, and environmental problems are strongly linked in a complex nexus that hinders development and threatens the region's food security, health, and natural resources. And just as the problems are connected, their solutions need to be integrated and mutually reinforcing to reverse this downward spiral.

For that reason, the importance of this Forum cannot be overemphasized. The stakes are high. They are nothing less than a sustainable future for humanity.

Two and a half years ago, heads of states attending the Earth Summit in Rio called for urgent action on sustainable development. Yet 30 months later, the primary message of the Earth Summit – the urgent call for action – has been obscured by economic recession, fratricidal conflicts, natural and man-made disasters, and escalating poverty. Moreover, while the Earth Summit's Agenda 21 acknowledged the relationship between population, consumption, and natural resources, it failed to adopt the policies needed to address population growth and development. Now the challenge is to integrate UNCED with the outcomes of the United Nations' Cairo population conference. But we can redeem ourselves. A forum such as this one today that is being organised by the United Nations University and sponsored by

Obayashi Corporation provides us with the opportunity and the responsibility to answer that urgent call for action – and to advance the sustainable development agenda in a meaningful way.

Regardless of a country's level of development, population growth means increased energy use, increased resource consumption, and environmental stress. It cannot be clearer that we are draining our planet's ability to support ourselves.

For many of us living in developed countries, we are becoming well acquainted with population-related problems, whether we recognise them as such or not. Air and water pollution, difficulties in siting landfills, loss of migratory fish stocks, congested roads and waste in our cities are just symptoms of the problem of population growth.

For those living in developing countries, population growth affects the environment and quality of life in very basic ways. Rapid population growth increases pressure on resources, often forcing communities into unsustainable practices for the simple purpose of obtaining the food, fuel, and shelter needed to survive – literally forcing people into acts of environmental degradation.

Clearly, it is not an easy job to shape reasoned and responsible solutions to the most pressing problems of our time – population and consumption. The United Nations has devoted a series of conferences to this subject. This road actually started in Rio with the Earth Summit in 1992; from there it went in September last year to Cairo for the International Conference on Population and Development. The next stop was Copenhagen for the World Summit on Social Development last March; then it is on to Beijing for the World Conference on Women in September this year, and finally to Istanbul in June 1996 with a City Summit called Habitat II Conference.

This unprecedented continuum of conferences spans some of the most serious and pressing challenges of human security that will confront the world community in the new century. Cumulatively, the conferences already held and those still to take place have begun adopting a more holistic and humane approach towards our global problems, and towards the cooperative solutions they require.

The effects and endeavours of the United Nations University fit well into this concept. Its Programme on Environmentally Sustainable Development, the Zero Emissions Research Initiative, and the Global Environmental Fora – all focus on global problems which require a vision and concerted effort to overcome. If overpopulation is, indeed, our greatest overall problem, then surely research, education, and training must be our most effective long-term weapons against it, for the elimination of deprivation and ignorance.

The United Nations Environment Programme's International Environmental Technology Centre – based here in Osaka and also in Shiga Prefecture – would like to join forces with UNU and the experts assembled at this Forum to find answers on how to ensure the sustainability of development processes. The answers do not lie in an "end-to-development" philosophy, not in restrictions of the legitimate aspirations of developing nations, not through coercive policies or limitation of individual choice, and not in deepening ideological divisions.

Whether we focus on controlling unchecked rise in population growth or the unsustainable trends in land management or the resulting environmental changes, what is required is:
- a global commitment to action;
- the assumption of individual responsibility; and
- a return to the original meaning of the word "development" – meaning the unfolding of potential.

We at the International Environmental Technology Centre know that unfolding this potential means to address not only the scientific, technological, and economic aspects of global development problems, but also the human, social, and legislative issues involved. We try, bearing all these aspects in mind, to promote the utilisation of environmentally sound technologies in the hope of helping to create a more equitable and less wasteful society. I believe that the experience we have gained in Japan, in North America or western Europe through our own industrialisation histories, and the appropriate technology we have been able to develop in the process, may be of real assistance to developing countries and countries with economies in transition. Because mismanagement in our past world of modest population and limited technological power could be tolerated far more readily than it can in the crowded, high-tech global megalopolis of tomorrow, we have to do better, we urgently need new thinking.

It only remains for me to thank you all for your patience in listening to me. I wish the UNU Global Environmental Forum number IV every success, and I am convinced it will make a most valuable contribution to reversing the disingenuous spiral of population–environment linkages towards the direction of sustainability.

1

People, Land Management, and Environmental Change: The Problems That a United Nations University Programme Is Studying

Harold Brookfield

Background

The project described in this introductory paper focuses its attention on one specific part of the whole complex relationship between people, their society, and the land. We concentrate on small-farm regions of the tropics and sub-tropics and, within this, on how farmers manage or mismanage their land and its biological diversity, and on why they do so. Some such areas are well managed, and some suffer severe degradation. Dealing with small-farm regions, we encounter great diversity of farming practices, often within quite small areas, and we term this "agrodiversity" (Brookfield and Padoch 1994). Although PLEC (People, Land Management, and Environmental Change) includes some experimental research based on farmers' practices, it draws information principally from what farmers do and know, and from its own field research. PLEC is also a capacity-building project, with a main objective of creating a multidisciplinary network of specialists in environmental management, qualified also to understand the societal forces which create the conditions within which small farmers manage their land and biota.

We do not try to cover the whole of the tropics and subtropics, but instead concentrate our efforts on a few areas where these problems can be studied in greater depth. Six "clusters" of PLEC scientists have been formed, based in universities or government research organizations, in Africa, Asia-Pacific, and tropical American regions. They are linked together by the core project network. Four clusters work only in humid areas but two, in West and East Africa, extend into the dry subhumid margins. In mid-1995 there are 70 scientists in PLEC, 53 of them working for institutions in the developing

1

countries. The distribution of the clusters, and the areas in which they work, is shown in figure 1. A good deal more about some of these clusters and their work is presented in the following papers.

To explain better what we are doing, we sometimes use supplementary titles for our project. One such is "Agrodiversity, Land Management, and Biodiversity." We argue that the greater part of the conservation problem, whether of land or biodiversity, lies in agricultural areas where crisis conditions have widely been generated by population growth, commercialization, deforestation, and land degradation. We urge a positive rather than a negative view of this situation, and propose that one of the most undervalued elements is the ability of a significant proportion of the world's small farmers to modify their farming systems in a comparatively short period of time, often in ways that make sustainable use of biodiversity as well as the land. Their object is production and not conservation, but an objective of sustained production does call for conservationist methods, and many farmers have or acquire the knowledge to adopt such methods and do. Going further, we also seek to persuade that farmers' adaptations and the knowledge on which they are based could often be successful, if societal conditions permit and assist. People manage the land and its biological diversity; moreover, they suffer the consequences of wrong decisions. One principal objective of PLEC is to give farmers' knowledge and practices an equal place with the findings of external science in the search for sustainability with conservation.

This paper does not describe the work of the project, though some of this is used, but discusses a part of the background. This paper looks at the specific question of intensification and innovation in land-resource management. It emphasizes some recent African material. PLEC argues that farmers can and do change their systems of management when pressures arise, and there is already abundant evidence to support this from within the project, as well as from a wider literature. My aim is to look outside PLEC's own evidence, and put it into a wider context. To discuss this question, I draw initially on two papers which I wrote about a decade ago, and which are here reconsidered (Brookfield 1984, 1986). This short paper is part of a rethinking process that I hope to develop further in other writing now under way. The reconsideration presented here is in the light of some recent literature concerning the way in which land management has adapted, with considerable success, to increase in population and to the need to manage land degradation. Some of this adaptation constitutes reclamation of degraded land and this aspect, not stressed in the mid-1980s' papers, is given greater weight here.

Intensification and Innovation

To begin, it is necessary to define some basic terms. *Intensification* is properly measured against constant land, and means measures which will enhance

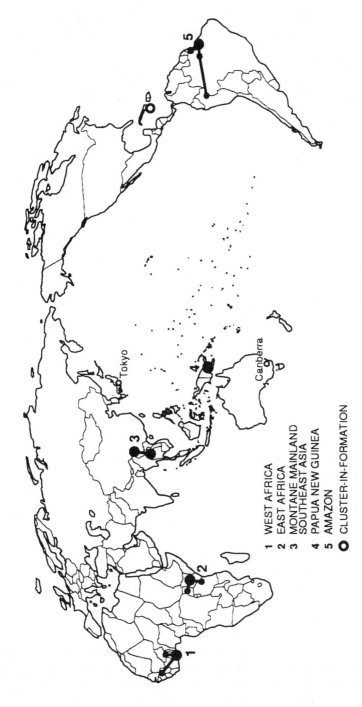

1 WEST AFRICA
2 EAST AFRICA
3 MONTANE MAINLAND
   SOUTHEAST ASIA
4 PAPUA NEW GUINEA
5 AMAZON
○ CLUSTER-IN-FORMATION

Figure 1 The Clusters of PLEC

productivity on that land. More specifically, what is usually implied is the intensification of inputs, meaning their increase on constant land up to – or even beyond – the point of zero marginal return. In most of the literature the inputs that are discussed are those of labour, since increases in capital input are generally "lumpy," and less easy to analyse. However, where changes are introduced into an agricultural system the inputs are applied in qualitatively new ways, and a new "curve" of intensification is originated. Such changes are best separated from progressive intensification, and thought of as *innovations*.

Innovations, in turn, need to be distinguished into those which are specific to a particular crop or crop cycle, and those of greater durability. The latter may be purely organizational, such as a change in land tenure arrangements, although these normally accompany changes in the farming system. A major set of innovations takes place when, for example, a mainly pastoral system becomes dominantly agricultural. Certain innovations create capital in the land, often termed "landesque capital." Mounds, ridges, terraces, irrigation and drainage systems, dams, created land and created soil, agroforests, fishponds and the like, constitute such capital. Their physical evidence may remain on the land after abandonment of upkeep, and some relinquished innovations may remain capable of being brought back into service after many years. Most successful physical innovations have the effect of changing the conditions under which soil fertility, structure, and soil water are managed, or of reducing or minimizing erosion from sloping land. They are changes in the direction of greater sustainability of production.

All innovations are not equal. Some are of only a minor nature, though they may be important. An example might be the deliberate fostering of a particular tree species as a part of general fallow management, improving nutrient recycling along with other benefits. Another might be the introduction of row-planting of rice in place of more random planting in a wet field, to facilitate weeding with higher planting density. Major innovations, however, include the terracing of hillsides, the introduction of a system of irrigation, the placement of mulch in mounds or under ridges, or anything which creates a major change in the system of management. The distinction is easier to draw conceptually than in practice but it is useful because major innovations are those which mark a real change in the conditions of production and, often, the creation of enduring landesque capital.

The relationship of major to minor innovation and of both to intensification is fairly easily understood in the case of wetlands. Wetlands may be agriculturally usable when water levels are low without any special works at all, but crops are at risk from unexpected variations in water level. Production is insecure. Once the wetlands are drained, or ridged, however, dry land is created. The effect of the major innovation is an immediate boost in crop security and often in per-crop yield as well. Such a system may then be elaborated with new minor innovations such as varying the height of the ridges, and working them more intensively by applying more labour. The initial transformation of the land surface was, however, the major innovation, greatly changing the conditions of production. There may be further major innovations, for example involving management of the water level. However, most subsequent elaborations are of a minor order, accompanying intensification of inputs into the new system.

It is common to see major innovations on difficult sites that could not have been managed sustainably without them. Wetlands and steeplands can both be used without innovation, and so can difficult soils, but their continued productivity cannot often be achieved without some form of physical transformation of the site. There are few steep hillsides that can be worked for a long period without loss of soil, unless there is some form of slope protection. While on naturally resilient slopes the problem may be considerably less than many observers, seeking to introduce high-technology management, may have believed, there are always advantages in the introduction of protective measures. However, it is not only in such obvious locations that innovations are important. Where land has been cleared from forest, for example, considerable changes take place in its micro-climate and, where the general climate is characterized by drought, the effect of this drought may be greatly intensified on treeless land. Innovations both to adapt agriculture for greater drought tolerance, and to restore some tree cover, may therefore be highly advantageous.

Rarely are innovations introduced all at once, unless by major external intervention. Both logical reasoning and observed experience show that they demand farmers' experimentation, either spontaneous or using new information they have received. Sustained and widespread adoption only takes place after clear advantages have been demonstrated. So little has the adaptive capacity of farmers been valued during the past two or three generations of development research and intervention, that the view of farmers as experimenters has until very recently indeed been derided, and it is still derided in some areas. Yet farmers experiment constantly, with new crops, new trees, new tools and practices, and new ideas. Many experiments, even if successful, are allowed to lapse, but a proportion survive and are more widely adopted. It is this proportion of enduring innovations, repeated across region after region, that is responsible for the enormous diversity of farming systems found across the world.

There is, however, a problem with some systems which involve a large inheritance of landesque capital from an earlier period of time, even a very recent period. The maintenance of this capital can be an onerous burden, absorbing a substantial amount of labour input, and income, that cannot therefore be devoted to current production and income-gaining. Modern change usually includes substantial growth of off-farm employment, whether or not it involves emigration from rural areas. The employment is often selective of young people, and the farm population grows older. The effect of a heavy dependence on landesque capital is most obvious in the intensive hydraulic systems of Asia, with their heavy dependence on irrigation, terracing, and their works. Writing of China, but with much wider potential application, Elvin (1993/4) writes of "technological lock-in," which can lead swiftly from apparent sustainability to unsustainability. While this is not yet a problem in the specific cases discussed below, it is already there as a threat.

One of the reasons why farmers' experiments have been so little recognized is that most external observers remain in a farming region only from hours to, at most, a very short period of years. Yet change in farming systems takes place across periods of decades and it is only now, when modern observations can be set against those of from 20 to 60 years ago, that the amount of largely spontaneous change can clearly be seen. It could have been realized sooner, by logical thinking about what must have taken place to produce present management systems in the many areas where whole new crop complexes have been introduced by transfers between continents within the past one to five hundred years. Except in an academic literature, however, the necessity to accept farmers' experimentation as a reality did not arise. And a good deal of the academic literature was also put aside as irrelevant by the practitioners of "development," or was used only in a highly selective fashion. One of PLEC's secondary objectives is to strive for larger incorporation of the results of good research into diagnosis, management, and development practice.

A Case in East Africa

A useful body of literature has emerged in recent years on the modification of African agriculture, principally in the face of rising population pressure. In what follows, I concentrate heavily on a small part of this literature. A considered official view is also the classic view that, while African farmers may have the ability to change their farming systems, the process is too slow to cope with the rate of contemporary population growth (Cleaver and Schreiber 1992). A recent group of academic studies leads to a less negative conclusion, noting a considerable amount of successful adaptation, but leaves the case regarding population growth open (Turner et al. 1993). Several of its contributors still regard it as "inevitable" that continued growth of population will lead ultimately to land degradation. Very different is the basic conclusion of a detailed study of change in the Machakos district of Kenya, which concludes that increasing density of population over most of the present century has been a major force for improved environmental management (Tiffen and Mortimore 1992; Tiffen et al. 1994). Specifically, increased demand and closer interaction generated by a denser population, coupled with increased labour supply and economies of scale, have more than outweighed the negative effects of population growth. After a six-fold increase in population over 60 years, dryland farming in a drought-prone subhumid region "has been turned toward a new logic of sustainability" (Tiffen et al. 1994, p. 12).

There is more to the Machakos case than this, however. The processes of change involved first a progressive, indeed rapid, degradation which endured through the 1940s and later in some areas. Its most obvious, and best described, characteristic was soil erosion, but deforestation, declining soil fertility, and declining availability of water were also important elements. Already in the 1930s, however, improvement was recognizable in some areas, accompanied by a shift from largely pastoral farming supplemented by rotational cultivation toward mixed farming in the wetter areas, and with the beginnings of terracing and water control. External intervention was strongly directive and was concerned primarily with soil conservation, a pattern which continued until the 1950s. The imposed innovations were only selectively accepted by the farmers. Their family structure was at the same time shifting from extended toward nuclear families, more closely identifying the individual household with a specific tract of owned land, so that restorative management arose increasingly at the family-farm level, and not over whole landscapes. From the 1950s onward colonial government discouragement of cash cropping was replaced by encouragement of an already strong trend toward planting of coffee and other commercial crops and, coupled with incomes earned off the farm, cash crops began to provide money capital for investment on the land. Top-down intervention began to be replaced by persuasive agricultural extension by a service which continued to be strong until the end of the 1970s.

Positive change accelerated in the 1960s and has continued. It has included extensive terracing by different methods adapted to soil and slope conditions, cut-off drains to divert and conserve water, heavy though still insufficient use of livestock manure on the farms, adoption of a suitable plough, cash cropping of growing diversity, substantial increase in off-farm employment and business and, almost wholly by indigenous innovation with only highly selective adoption of external advice, the evolution of complex systems of agroforestry (Rochelau et al. 1993). By the 1990s, steeplands in central Machakos which were bare and eroded in the 1930s are well treed, intensively terraced, and managed by a diverse range of strategies. However, there are still clearly visible contrasts between individual holdings in the application of skills and labour to sustainable management.

Erosion is greatly reduced, even in the drier lower regions, though it has not ceased. Decline of soil fertility, due to soil loss and inadequate replacement of nutrients, also continues, although far less on the better-managed farms. Tree cover has increased almost everywhere, and in the more densely peopled areas it consists almost entirely of useful species, a high proportion of them exotics. Densities in some areas are now above 400/km^2, and population is still increasing at a high rate, but the on-farm density is now growing less fast because of the growth of off-farm means of gaining a living. However, land is now very scarce, encouraging the adoption of all possible methods to raise family-farm productivity. The net effect is an increase in per-capita production and incomes, and a great improvement in prospects for the future. All this has been achieved by a fruitful mix of externally derived and endogenous innovations, large and small, and the pattern of progressive change has not ceased.

Machakos is exceptional, but not unique. There are areas in Kenya and elsewhere in East Africa where management is yet more intensive, supporting still higher population densities in a sustainable manner (R.M. Kiome pers. comm.). But there are other areas in which management continues to leave much to be desired. Machakos has some special characteristics, including its proximity to the national capital with its large consuming population, as well as a good internal infrastructure. None the less, there are some important lessons of wider application to be derived. While externally derived innovations have been of major importance, they have been filtered through farmers' own perceptions of rationality, and several major innovations have been adopted without official encouragement. The most important lesson is that the capabilities of indigenous farmers should not continue to be undervalued, as they have been in the past.

The Management of Degradation: A West African Case

By a wholly realistic interpretation, much of what has taken place in Machakos is the reclamation of a degraded landscape. There are large areas of degraded land in Africa as in other regions of the world. In many of them the fruitful combination of external intervention with spontaneous innovation which has characterized most of the Machakos story is absent. Government intervention, like international advice and aid, has been heavily oriented toward the production of commodities for the export market, and little assistance is given to farmers who lack the conditions for production of such commodities. The late-colonial concern with soil conservation, which was the true origin of most of the useful direct interventions in the Machakos case, was never strong where soil erosion was not a visible problem. Concern with deforestation, active in the post-colonial period and very popular today, has in most cases not got beyond the stage of unenforceable prohibition of clearance and use of fire.

Farmers' capability, and willingness to experiment, are however widespread elements in the situation. A growing literature is now emphasising this fact for policy makers, but it is very recent, and the pioneering writing of Paul Richards in West Africa is not yet overtaken (Richards 1983a, 1983b, 1985). If farmers have a well-integrated system of resource management developed in the past, and still evolving, they can make successful adaptations to change, even enhancing the quality of landscape and biodiversity. Such a case is described in Guinea, West Africa, by Leach and Fairhead (1994, 1995). Valley swamps have been managed and enlarged for wet rice, large woodlands within which forest species have also thrived have been planted around villages, and the savanna has been mulch-mounded for cassava and has become more woody in consequence. Population growth has been accompanied by regression of the savanna, not by its enlargement. This conclusion runs in the face of still-widespread belief that savanna is everywhere extending at the expense of forest. Yet the conclusion is incontrovertibly supported by the comparative use of old and modern air photographs, supplemented by remote sensing imagery, and by descriptions found in the archives and in the memories of the farmers and their families.

Although there are also parts of West Africa in which intensive farming in subhumid areas has been notably successful, it does not seem that this is the case in much of Ghana. Large areas of semi-deciduous forest, once underplanted for cocoa and oil palm, have now become "savannized" and farmers experience declining yields, diminished availability of firewood, and inability to cultivate the more demanding crops that were common only one or two generations ago. Government support in both colonial and post-colonial times has strongly emphasized export cash crops, and continues to do so. Farmers face land degradation largely on their own.

Two recent studies by PLEC members, in adjacent areas of south-eastern Ghana, present somewhat differing interpretations. This is a region across which waves of pioneering land settlement have passed during some two centuries, almost from the outset strongly associated with production for the world market. Only the most naturally resilient parts of Ghana's original oil palm and cocoa heartlands remain in good productive condition, now for food crops grown for the urban market. Other areas, penetrated first for oil palm and then for cocoa farming, mainly in the present century, now suffer widespread deforestation, biodiversity loss, and impoverishment of soils. In the Mampong valley and adjacent areas, Gyasi et al. (1995) find farmers' attempts to modify cropping patterns and intensifying production thwarted by poverty and technological constraints, and by the high proportion of tenancy among the farmers which inhibits innovation and encourages overexploitation. Commercial pressures, and rising population pressure, create a potentially critical situation.

In the adjacent Manya Krobo district, Amanor (1994) encounters many of the same phenomena and a very similar history, but also sees more positive elements. Tree selection is of major importance, as is the introduction of leguminous crops. Perhaps looking more specifically for innovation, Amanor finds it among a minority of farmers who introduce comprehensive management of the fallow so as to obtain better crops after it, or have developed an agroforestry system based on a particular tree (*Newbouldia laevis*). This tree has long been known to provide the right amount of shade for undercropping and to create a more favourable soil environment. Farmers' agroforestry treatments, spontaneous but very similar to those advocated by external specialists for use with introduced trees and shrubs, further enhance this environment for crops. A number of other experiments, by individual farmers, were uncovered during field research. Furthermore, the main centres of innovation appear to be in the most degraded parts of the region, where established production systems work least well, where there is longer experience of degradation and perhaps where the remaining forest stock is reduced to its hardiest species, assisting farmers to harness the remaining energies of the forest system. Over quite a small area, considerable variety has evolved out of an earlier greater uniformity of agroecosystem type. None the less, the management of degradation is still only marginal in the region as a whole.

A Case of Stagnation in Papua New Guinea

The author has worked for more than a decade with people and their agriculture in the highlands of Papua New Guinea, in a densely peopled area known as Chimbu where an elaborate farming system, complex and adaptive to different ecological conditions, was developed over some two hundred years in a pre-colonial period that ended only in the 1930s. With return visits, he has data on land use extending over a 26-year period from 1958 to 1984, with subsequent observation up to 1990 (Brown et al. 1990). During this period the population of the area specifically studied has grown by more than 50 per cent, but innovation tailed off rapidly after the first years of observation. During those early years coffee was introduced as a cash crop and by 1960 had already been concentrated on the most suitable land, while the regionally ubiquitous casuarina tree, which fixes nitrogen, had quickly replaced introduced exotic shrubs as shade. A number of related changes had been made in the farming system, though hindsight shows these to have been related as much to the beginning of concentration of settlement into quasi-villages, with family residence taking the place of gender separation, as to the coffee innovation itself. After the mid-1960s, however, coffee development reached a plateau, other new crops came and went, while the basic farming system scarcely changed at all. All the population conditions that led to innovation in Machakos were present, but innovation in Chimbu virtually ceased.

The stagnation has social and political causes, for this has become a depressed region. However, two additional conditions which were present in Machakos have been absent in Chimbu. A very sustainable set of land management adaptations was already developed in pre-colonial times and, moreover, they continue to give successful results notwithstanding the rising density of population. While desire for higher incomes, and greater participation in economic opportunity, have been effective in bringing about change in some other parts of Papua New Guinea, in Chimbu there has been an absence of sustained opportunity, while there has also been no significant land degradation in modern times to act as a spur. This is of interest in the general context, because it supports a view that population growth is not a sufficient cause of either land degradation, or improved land management – according to which theoretical position one adopts. Other elements need to be present in all cases.

Discussion and Conclusion

Diverse management of diverse landscapes by small farmers is more than an inheritance from the past. It continues, and evolves. But adaptive diversity is also at risk, for both modern science and the pressures – and attractions – of commercialization have acted to reduce agrodiversity and replace it with simple, widely replicated systems of farming. Often, these modern systems are destructive of the biodiversity sustained under agrodiversity, not least the biodiversity of the soil. They may, at least for a time, support more production, but they manage the land less well. PLEC seeks to understand and explain these small-farmer adaptations, whether they are "traditional" or not. Indeed, the adaptations rather than the old systems themselves are our *main* focus of interest, for these adaptations are dynamic and responsive to change in external conditions. And, because we study change, we are also concerned with the history of change, in agriculture, society, and environment alike. Although many have tried to do it, conclusions about change cannot be established without studying history.

There are other elements in our project, some of which are mentioned in other papers in the collection, but those discussed here are among our distinctive characteristics. We emphasise the study of agrodiversity and its value, the study of adaptation and change, and the history of change. With them, and through our clusters of scientists spread across the tropical world, we seek, by building a capacity for informed, action-oriented research, to join a much larger endeavour, that of finding paths to a sustainable future for the growing numbers of people in the developing lands.

REFERENCES

Amanor, K.S. 1994. *The New Frontier: Farmers' Response to Land Degradation. A West African Study.* Zed Books, London.

Brookfield, H. 1984. Intensification revisited. *Pacific Viewpoint* 25, 15–44.

Brookfield, H. 1986. Intensification intensified: "Prehistoric Intensive Agriculture in the Tropics" (review article). *Archaeology in Oceania* 21:3, 177–181.

Brookfield, H. and C. Padoch. 1994. Appreciating agrodiversity: A look at the dynamism and diversity of indigenous farming practices. *Environment* 36:5, 6–11; 37–45.

Brown, P., H. Brookfield and R. Grau. 1990. Land tenure and transfer in Chimbu, Papua New Guinea: 1958–1984 – A study in continuity and change, accommodation and opportunism. *Human Ecology* 18:1, 21–49.

Cleaver, K.M. and G.A. Schreiber. 1992. *The Population, Agriculture and Environment Nexus in Sub-Saharan Africa.* The World Bank, Africa Region, Washington, DC.

Elvin, M. 1994. Three thousand years of unsustainable growth: China's environment from archaic times to the present. *East Asian History* 6 [1993, publ. 1994], 7–46.

Gyasi, E.A., G.T. Agyepong, E. Ardayfio-Schandorf, L. Enu-Kwesi, J.S. Nabila and

E. Owusu-Bennoah. 1995. Production pressure and environmental change in the forest-savanna zone of southern Ghana. *Global Environmental Change: Human and Policy Dimensions* 5:4, 355–366.

Leach, M. and J. Fairhead. 1994. *The Forest Islands of Kissidougou: Social Dynamics of Environmental Change in West Africa's Forest-Savanna Mosaic* (with D. Millimounou and M. Kamano). Report to ESCOR of the Overseas Development Administration. Institute of Development Studies, Brighton, UK.

Leach, M. and J. Fairhead. 1995. Reading forest history backwards: The interaction of policy and local land use in Guinea's forest-savanna mosaic, 1893–1993. *Environment and History* 1:55–91.

Richards, P.W. 1983a. Farming systems in West Africa. *Progress in Human Geography* 7, 1–39.

Richards, P.W. 1983b. Ecological change and the politics of African land use. *African Studies Review* 26:2, 1–72.

Richards, P.W. 1985. *Indigenous Agricultural Revolution: Ecology and Food Production in West Africa*. Hutchinson, London.

Rochelau, D., K. Wachira, L. Malaret and B.M. Wanjohi. 1993. Local knowledge for agroforestry and native plants. In: R. Chambers, A. Pacey and L.A. Thrupp (eds.), *Farmer First: Farmer Innovation and Agricultural Research*, 14–24. Intermediate Technology Publications, London.

Tiffen, M. and M. Mortimore. 1992. Environment, population growth and production intensity: A case study of Machakos. *Development Policy Review* 10, 359–387.

Tiffen, M., M. Mortimore and F. Gichuki. 1994. *More People, Less Erosion: Environmental Recovery in Kenya*. John Wiley, Chichester, UK.

Turner II, B.L., G. Hyden and R. Kates (eds.). 1993. *Population Growth and Agricultural Change in Africa*. University Press of Florida, Gainesville.

2

Land Management for Sustainable Development: Farmers' Participation

Michael Stocking

Introduction

An Anecdote

Shinyanga Region in northern Tanzania must be one of the most degraded parts of a degraded continent. Africa is often perceived as having the worst excesses of environmental destruction – desertification; forest removal; soil-nutrient and organic-matter depletion; and, above all, soil erosion with its attendant bare hills and incised gullies. Although intrinsically fertile because of the calcium-rich lake-bed sediments, the south-eastern parts of Shinyanga manifest the full range of indicators of environmental stress. Since the implementation of *Ujamaa*, or villagisation policy, in 1972/73, vast areas of *Acacia–Commiphora* woodland have been burnt to make way for extensive cultivation of cash crops such as cotton and subsistence crops of maize and millet. The WaSukuma brought with them huge herds of cattle which rendered the seasonal grasslands almost totally bare. In a few short years, the effect of the population transition was devastating.

Yet, amidst the sheet-eroded agricultural slopes and bare rangeland, our 1980/81 surveys for the World Bank–funded Regional Integrated Development Programme (EcoSystems 1982) found pockets of hope: places where land use seemed far less exploitative; where intricate systems of small-scale agriculture were producing a variety of crops; where locally available technology (animal-drawn ploughs) had been adapted to build contour ridges; where intercropping on the ridges enabled good conservation of soil and water as well as provision of cereal crops, vegetables, legumes, and fodder for animals; where the social system controlled indiscriminate grazing by animals. Ironically, these very places were generally isolated, had not been visited by agricultural officers or extension services, and were on some of the poorest soils.

Participatory Land-Management Issues

What was going on in these remote spots of Shinyanga Region? The answer embraces a number of issues which have come to the fore in recent professional understanding of the importance of farmers' own knowledge and perceptions of their environment, as well as their development of solutions and technologies of natural resource management. The issues are not solely about what scientists and traditional science may contribute. As Brookfield (1993) has remarked about experts who gathered for a UNU conference on "Sustainable Environmental Futures for South-East Asia" they find it much easier to say what is *unsustainable* about present resource use, than to identify what might be *sustainable*. Scientists blame policy makers, the ineffectiveness of development efforts, rapid population growth, and inadequate social and economic systems – indeed, anything but themselves. This paper is an attempt to redress the balance by emphasising ingredients which promote sustainable land management and to outline an approach which values non-traditional scientific knowledge as of equal or greater worth than the normal offerings of reductionist science.

Specifically, this paper adopts three positive and, hopefully, optimistic themes:

• Sustainable land management is possible in difficult environments – it is not inevitable that marginalised, small-scale farmers will ruin their natural resources for short-term gain.
• Societies have the capacity to adapt and to change to new circumstances – they do this by developing new ways of meeting fuel, fibre, food, and fodder needs, by adapting indigenous and exogenous technologies, and by using their own informal experimentation.
• Farmers often have the solution to their own problems – solutions which may differ from those promoted by external agents; the role of the professional is now evolving into the provider of assistance to help unlock indigenous capability.

The logical conclusion to these three themes is that farmer participation is essential, even mandatory, for effective rural development and the sustainable management of land and water resources. The paper will conclude with the key components of a farmer-participatory approach to natural resource management.

Sustainable Land Management is Possible

Sustainable Land-Management Practices

There is an increasing record of sustainable land-management practices in an array of environments which had been seen as difficult, marginal, and scientifically challenging. In a major new review of policies and practices for sustainable agriculture, Pretty (1995) describes a large number of resource-conserving technologies, many of which were either discovered by farmers or developed in partnership between agricultural research and local people. Some happened by accident; others were deliberate and planned interventions. Similarly, in soil and water conservation, the literature which is recognising the value of some indigenous technologies (e.g. IFAD 1992) and of new directions in land husbandry and working with farmers (e.g. Lundgren et al. 1993) is flourishing. The list is impressive and could include:

- *integrated pest management*; now widely used in the USA, but has greatest potential application in developing countries through using local and scientific knowledge on resistant varieties, alternative natural pesticides, bacterial and viral pesticides, use of pheromones, appropriate rotations and multiple cropping, and habitats for natural enemies.
- *integrated plant nutrition*; with an emphasis on improving efficiency of uptake of applied nutrients, introducing beneficial crops, and using organic sources. Some of the most spectacular developments have been in the use of legumes as part of an overall soil management, weed control, minimum tillage, and plant nutrition strategy – see below.
- *agroforestry*; where woody species are combined with land uses such as arable and rangeland. An interesting example cited by Palmer (1992) is where farmers have taken particular components of the Sloping Agricultural Land Technology (SALT) model of contour hedgerows which has been widely promoted in the Philippines and Sri Lanka, and adapted them to their own systems. The pure SALT seems to be consistently ignored unless subsidies and inducements are provided.
- *sediment traps*; built in a gully often as an earth embankment, these trap sediment from upslope, creating new fields and additional production opportunities. In Rajasthan (India) the plugging of *nullahs* has been practised for decades, but has not been recognised as a valid practice by the professionals – and hence has received no state aid. Yet, farmers willingly and voluntarily carry out the labour because they perceive the practice to give them a more valuable and reliable crop than any other land-management practice. These and other soil and water conservation practices are only recently being described and catalogued (Kerr and Sanghi 1992) – an example is given later in this paper.
- *water harvesting*; in semi-arid zones, the concentration of water to provide more effective use of rainfall to parts of the landscape is widely accepted and many techniques exist (Critchley and Sievert 1991). For example, "fish-scale terraces" in China provide minicatchments of a square metre or more to water a single tree on each "scale" on steeply sloping land.

Example: Velvet Bean in Brazil

To illustrate further the feasibility of sustainable land-management practices in difficult, marginal environments, take the almost ubiquitous use of *Mucuna pruriens*, or velvet bean, by small farmers in the western portion of the southern Brazilian state of Santa Catarina. Spontaneously, the use of *Mucuna* has spread from farmer to farmer on the nutrient-poor oxisols which are intensively farmed on small plots for maize.

Briefly, the agricultural cycle is: (1) in spring, the hand-planting of maize with a dibber into a thick layer of dead *Mucuna* mulch – no need for ploughing as weed suppression is achieved by the mulch and microfaunal activity in the soil has kept the soil aerated; (2) the maize germinates and commences growth, assisted by additional water-holding capacity, the humidity, and extra nutrients of the mulch; (3) later in the growing season, self-sown velvet-bean seeds also germinate – these are largely non-competitive with the maize and achieve maximum crop-water demand only after the maize has matured; (4) the maize is harvested and the *Mucuna* continues to grow on residual soil moisture; (5) when winter temperatures return, the *Mucuna* dies back, leaving fixed nitrogen in the soil and good surface-soil protection against erosion. Adapted by farmers but with the original species introduction attributed to formal research systems, the benefits of using *Mucuna pruriens* may be summarised as: reduced labour in land preparation, weeding and planting; low-cost and no-cost inputs of nitrogen; water conservation through the use of mulch; soil conservation through excellent ground cover of both maize and velvet bean; and good returns to the farmer estimated as up to five tonnes per hectare of maize with little need for expensive inputs.

Resilience and Sensitivity

Sustainable land management must, however, depend on the intrinsic quality of the natural resource base. Technologies, even farmer-developed ones, cannot be made to work everywhere and under all conditions. Environments are difficult and marginal precisely because their use is constrained and options for exploitation are limited. A useful means of distinguishing environments is provided by the conceptual terms "sensitivity" and "resilience."

Whole landscapes are now becoming increasingly recognised for their differences in sensitivity (Thomas and Allison 1993). The term implies the degree of fragility; how readily change occurs, especially with only small differences in external force; and the susceptibility of landscape components such as the soils to alter irreversibly. Landscapes have inherent sensitivities, but of more relevance to sustainable development is the likely effects of human activity and whether these change the attributes of the environment so as to undermine future land uses. Sensitive environments degrade easily. For example, the Loess Plateau in China erodes at an apparently alarming rate with only modest changes in land use. The spectacular gullies and sediment-choked rivers are ample testimony to the sensitivity of the landscape system to human activity. Another example is provided by the luvisols (or alfisols) common in seasonal rainfall regimes in the tropics where much subsistence farming activity is carried out. With their concentrated nutrient distribution in the topsoil, luvisols lose their fertility very easily. The two examples are different, however, in their ability to cope with the changes and recuperate. That difference is encompassed in the term resilience.

Resilience is the property that allows a land system to absorb and utilise change. It is its resistance to external shocks and the resultant changes that are brought about. In the Loess Plateau example, the soils are resilient. They have good reserves of nutrients, and the societies that use them are adept at bench terracing, the use of irrigation, and fertility maintenance techniques. Thus, even with very high rates of erosion, the Loess Plateau remains relatively productive. From fieldwork in the Niehegou Catchment on the southern part of the Plateau, we have even found farmers who deliberately, they say, seek eroded slopes to grow potatoes because yields are better there. Consequently, one can call such landscape conditions highly sensitive but also highly resilient. In the second example, it is only with very great difficulty, expense, and effort that land management can restore the productivity of luvisols once they have been degraded – high sensitivity; low resilience. Similarly, many conditions have relatively low sensitivity to human interference and a high resilience to external shocks: the majority of temperate farming conditions are able to withstand intensive agriculture and the use of large numbers of inputs.

By defining land and natural resources in terms of their sensitivity and resilience, the various options for sustainable land management can be framed. This has recently been presented as a matrix for environmental managers seeking to classify the different properties of soils which make certain agro-ecosystems more hazardous than others (Stocking 1995b). By understanding the dynamic nature of our renewable natural resources through such concepts as resilience and sensitivity, the productive potential of the land should be able to be managed in a more sustainable fashion.

Societies Adapt and Change

Knowledge Systems

Cultures are often perceived as static, unchanging, unyielding, and inherently conservative. Stereotypically, local societies are seen as a barrier to development and modernisation; local people are characterised as a problem precisely because they will usually wish to cling to their existing ways; and local cultures are hidebound by myth and irrational practices. This view tends to be perpetuated by modern science, presented as a body of "fact," rational deduction, and the only basis for economic development. It is a science that is based upon high technology solutions, external inputs, and the results of experiments carried out under controlled and (usually) temperate conditions. Häusler (1995) has presented a useful schema for distinguishing the two types of knowledge:

• Western scientific knowledge (*episteme*): analytical, impersonal, universal, cerebral, logically deducted from self-evident principles, communicated in writing;

• indigenous knowledge (*techne*): based upon experience, personal, particular, intuitive, implicit, integral, orally communicated.

Under some conditions, interesting interactions may occur between the store of indigenous knowledge and the outputs of Western science that give land-use systems that are apparently sustainable ecologically and appreciated locally (c.f. Apfel-Marglin and Marglin 1990) – the *Mucuna pruriens* case from Brazil described earlier is one such beneficial utilisation of both knowledge systems. Conceptually, however, the two systems provide conflicting analyses and mutually contradictory results.[1] In its relations with developing countries and local cultures, science implicitly brands society as "backward," "primitive," and requiring "appropriate technology." This last term is often taken by developing country professionals as especially demeaning – it says that only simple technologies, perhaps technologies that have now been well superseded in the West, are to be promoted, with the implication that local people are too ignorant or incapable of coping with better.

Local Adaptations

What, however, is the evidence from local societies themselves when faced with change, whether brought about by uncontrollable factors such as drought or other sources such as conflict, population pressures, or economic crises? It would be wrong to say that peasant populations have all the answers; that they can cope with all pressures. Manifestly, they cannot. However, we do find a remarkable set of "grass-roots responses" in even the most marginal of societies. Take, for example, the social dynamics of groups vulnerable to drought in semi-arid areas (Barraclough 1995): pastoralists and agriculturalists resort to a variety of interesting and intricate strategies, even though many of the alternatives may look distinctly unattractive to them. These may include:
• adapting production and consumption patterns: pastoralists in the Sahel have developed sophisticated land-management systems in order to spread the risks associated with drought (Behnke and Scoones 1992). These systems include herd diversification and the maintenance of large numbers of so-called "low quality" animals. It also includes the use of animal and vegetation indicators in the use of natural rangeland. Much of this local knowledge would be unrecognised and rejected by modern science, yet its rationality is slowly being discovered by scientists in the last few years;
• adopting social structures to assist risk aversion: traditionally, in many pastoral communities, reciprocal obligations between different groups have become institutionalised in order to share risk. Common-property regimes are well known in this regard. In other cases, society has been propelled into private land ownership and intensification of crop production;
• finding alternative livelihoods: peasants can be extremely entrepreneurial in discovering new ways to seek a living. Sometimes, these ways are illegal:

growing drugs, smuggling, poaching, and distilling of spirits. But there are also many examples of diversification of activities into new crops, village-based industry, and added-value processing of plants and animals.

Example: Soil and Water Conservation, India

Soil and water conservation in the semi-arid parts of Dungarpur District, southern Rajasthan, India, provide a fascinating example of the difference between Western scientific knowledge and indigenous knowledge, and how local society has adapted to a degrading environment, increasing population, and limited livelihood options. This is a difficult physical environment where the most prized agricultural opportunities are in the bottom of valleys where pockets of soil can be retained by earth bunds (embankments) and water trapped to maximise plant–water availability. The bare slopes are used for grazing, and, perhaps inevitably, they are denuded of vegetation for all but the two or three peak months of the rainy season. The people of Dungarpur are predominantly tribals or Bhils, so far at the bottom of the social ladder that they do not even enter the caste system. A few may get labouring jobs in Gujarat, but most have to subsist from the local natural resources and their knowledge of how to eke a living from a reluctant environment. What would Western scientific knowledge advise for these people and their conditions? And what actually do the Bhils do?

The standard manuals of soil and water conservation, supported by such major programmes as the All-India Watershed Management Project, all promote catchment (watershed) protection through earth bunds constructed on the contour. Support is provided through subsidies for bund construction based on the volume of soil dug following the guidelines of the professionals as expounded in the official manual. Emphatically stated is that these bunds must be constructed first at the top of the catchment in order to keep as much soil and water on the slopes, and to protect the valley from eroded sediments. Then progressively farmers are advised to build further bunds down the catchment until the whole area is "protected." The scientific logic is inescapable. If our principal objective is to control erosion, then keeping soil as close to where it is formed, and water as close to where it drops, are rational means to achieve our goal.

The Bhils view their situation through a completely different optic. At the core of their concern is, naturally enough, the security of their livelihood. In a marginal environment, that security is necessarily short term: the vagaries of drought, insecure land tenure, the improbabilities of gaining off-farm income and so on mean that people must ensure production *now*. The longer-term view of, say, greening of the whole landscape is not really feasible or, indeed, advisable, for families who need a regular and as reliable as possible productive output from the local natural resources. Earth bunds are constructed at the bottom of the catchment. Most importantly these are not seen as ways of catchment protection but as sediment and water traps, a means of obtaining a new field and a new production opportunity. In such valley-bottom parcels of soil, rice may be grown if there is enough water, vegetables can also be produced, and, if there is sufficient residual water (likely, given the topographic position), a dry-season crop of pulses may further support their meagre living. By contrast, consider what the professionally recommended way would mean for production. Instead of a new field, or soil and water replenishment to an old field, the soil would remain on the slopes. These slopes are stony from decades of erosion; they are termed the "wastelands" with good reason because the surface is littered with gravel. Keeping the soil here will mean a little more grass will grow and last for a few weeks longer after the wet season. One or two animals extra may be grazed for two or three months on the slope. As one old man told my interpreter: "What is the point of protecting the slopes first, when what we really want is rice? Our animals can more easily eat the rice straw all year than a few extra bites of grass on these stones. And, besides, our earth bunds in the valley give us better grass anyway than these wastelands." The return on investment of labour into a standard catchment protection plan would be minimal compared to the additional production afforded by the ways suggested by local knowledge. Who, then, is right?

To answer this last question, it is informative to note that local people do follow the standard advice if they are subsidised to construct bunds on the typical piece-work rates given by the Indian government. Tribal women will queue for the opportunity to do soil and water conservation! Professionals will congratulate themselves that the message of soil and water conservation has at last got through. The women, however, are showing their adaptability in seeking alternative livelihoods – they are farming subsidies, instead of farming the land. Our field calculations suggest that, if they work reasonably hard, they can gain more by government subsidy in payment for their daily labour than they could get by alternative employment, including arable cultivation. It is little wonder, then, that most government-sponsored programmes of soil and water conservation are neglected after construction, and sometimes even deliberately destroyed. Local people will hope that next year a project will return to pay them to reconstruct the bunds! In complete contrast, our fieldwork in Dungarpur showed time and again that local farmers will construct bunds according to their local knowledge and own analysis of the situation without subsidy and willingly. They would, of course, be even happier if government paid them to do it that way also.

Farmer Participation is Essential

Farmers' Experiments

The priorities, assumptions, objectives, and modes of analysis of professionals will thus be very different from those of local farmers. We have argued that in many cases a knowledge system that derives from local experience may provide techniques of land management which are not only more acceptable but also protect the environment. Without the intimate

Table 1 Strengths and Weaknesses of Farmers' Experiments

STRENGTHS:
1. Focus of experiments on issues directly relevant to farm household.
2. Start with an accumulated knowledge of household members.
3. Experiments usually directed to improving the use of locally available resources.
4. Use evaluation criteria which are directly relevant to farm household: e.g. does the crop taste nice?
5. Observations are iterative; take place during routine farm work; and are generally continuous. It is a process of progressive learning.

WEAKNESSES:
1. Search for solutions based on limited scientific understanding of processes and limited technical knowledge of the options.
2. No control. Experiments may be over whole field, so no comparisons are possible.
3. Non-statistically justifiable conclusions. Bias and prejudice, as well as the experiment, may invade the results!
4. Causes may be attributed to simplistic reasons, and hence results cannot be extrapolated.
5. Dissemination of results imperfect; maybe limited by gender or to specific socio-economic groups.

involvement of farmers, such knowledge is unlikely to be accessed in any development or assistance programme. However, it is important to have a balanced attitude as to what local knowledge can provide. Table 1 presents a listing of the strengths and weaknesses of farmers' experiments, the major source of accumulated local knowledge which offers possibility of integration with scientific knowledge.

On balance, it is local knowledge and the results of farmers' informal experiments which should get highest priority. Why? Local farmers live more closely to the real problems, constraints, and opportunities afforded by the environment. Their experience will pick up complex interactions of variables which may be very specific to the local circumstance but which would have been rejected in scientific experimentation because of their local nature and the impossibility of researching every possible permutation of factor values (e.g. rocky, steep slopes; high rainfall; restricted growing season; cereals intercropped with legumes; no fertilizer; hand tools; family labour). Perhaps the most pressing reason to give local knowledge the first consideration is that, in the end, it is the local farmers who suffer from poor advice, failed technologies, and expensive solutions. Local knowledge will likely not make the great leap forward in production that, say, a new crop variety could do, but at least it is the lowest risk option. We, the experts, can just go back to our offices, and we are extremely adept at rationalising our failures; to the local farmer failure has more fundamental significance – starvation, for example – and cannot just be shrugged off.

Farmer Participation

If farmers' knowledge may be superior; if they are closer to the real local problems; if they can see things that experts might often miss; and if their

Table 2 Features of Participatory and Non-participatory Approaches

Participatory Approach	Non-participatory Approach
Participation	Instruction
Involvement	Observation
Learning-by-doing	Teacher says
Bottom-up	Top-down
Ownership	Employment
Empowerment	Obedience

objectives are more realistic for economic development – then farmer participation in planning, decision-making, implementation, and evaluation is absolutely essential. This is participation rather than just involvement. Table 2 contrasts some aspects of a particpatory approach with a more standard advisory approach to development.

The most controversial aspects of a farmer-participatory approach involve the notions of "ownership" and "empowerment," which in effect mean the transfer of rights of determination from professionals and traditional decision makers to local people. Governments feel uncomfortable; professionals feel humiliated; and local people may well feel perplexed. It must therefore also include confidence- and capacity-building, both in the wise use of transferred power and in the new role of professionals as facilitators.

There has now developed a whole new culture of "farmer-first" approaches, starting with the work of Robert Chambers on the concerns about the top-down nature of rural development (Chambers 1983), the development of this into an appreciation of what farmers can contribute in the field of agricultural research (Chambers et al. 1989), and most recently an exposition of how farmers' knowledge can be integrated with professional understanding in order to provide for secure livelihoods (Chambers 1993). Essentially, "farmer-first" involves the scientist as observer rather than manipulator; it calls for empathy with the conditions and needs of rural land users; it raises the status of individual decision-making by farmers above that of advice from the centre; it assumes that farmers' choices will be rational and also supportive of the collective good of society. Pretty (1995) has reviewed some 20 case-studies drawn from India, East Africa, and Central America which seem to show that the new culture is gaining ground and is showing positive results both in terms of crop yields and in professional attitudes towards farmers' knowledge.

Engendering Participation

One of the most fundamental changes in agricultural development in recent years has been the rapid expansion of participatory approaches which involve interactive learning between professionals and farmers. A growing list of methods now exist for (1) professionals to understand local people;

(2) local people to inform outsiders of their needs; and (3) local people to analyse their own conditions. As Pretty (1995, p. 174) notes: "The interactive involvement of many people in different institutional contexts has promoted innovation and ownership." He cites some common principles in the alternative systems of learning and action:

- A defined methodology and systematic learning process: all participants learn cumulatively from each other.
- Multiple perspectives: diversity is deliberately sought; bias and prejudice are not avoided; multiple descriptions of real-world activity are encouraged.
- Group-learning processes: group enquiry and interaction; different sectors and disciplines; outsiders and insiders.
- Facilitating experts and stakeholders: expert role is to help others achieve what they want.
- Sustained action: learning leads to debate about change and accommodation of conflicting viewpoints; debate seeks to motivate and strengthen capacity to remove obstacles and initiate action.

How can such laudable goals be achieved? Much can be learned from direct project experience. As Atampugre (1993) notes for the fascinating experiences with NGOs on the Projet Agro-forestier (PAF) at Yatenga, Burkina Faso, many of the best examples remain undocumented and handed down only anecdotally. However, that of course is the way that farmers' experience is accumulated. Nicholas Atampugre's brutally candid evaluation of PAF, *Behind the Stone Lines*, shows that there are no simple formulae for reaching, understanding, and motivating people to action. Inter-village conflicts, the tolerant attitude to corruption, and the turbulent social dynamics between project staff and local people all make the simple goals above rhetorically easy to say but difficult to achieve. Nevertheless, distilled from experiences such as these which have had undeniable technical impacts – in the Yatenga case, the digging of *diguettes* and other soil- and water-conservation measures – there is an accumulating fund of participatory methods for alternative systems of learning and action which have proved themselves for some groups in some situations. Table 3 lists a number of these methods.[2]

As practitioners have found, participation calls for collective analysis, and when it works well, groups can, through shared perceptions, make significant advances in designing developments and motivating to action. Important common elements will include the visual construction of a problem using local criteria and illustrative material. Sensitive, non-intimidatory interviewing is essential: several observers have described such interviews as being more like structured conversations. In the final analysis, however, participatory learning requires empathy on the part of practitioners and openness on the part of local people – in short, it sets a bond of trust, and that trust will need hours of dialogue to build. The 1980s paradigm of "Rapid Rural Appraisal" has had to give way to "Participatory Appraisal" simply because the process cannot be rushed. Partnerships take time.

Table 3 Participatory Methods of Rural Appraisal; Alternative Systems of Learning and Action

* Direct field observation
* Key informants ("local experts")
* Focus groups/group interviews
* Participant observation
* Transect walks (walking with the locals)
* Participatory mapping/village landscape modelling/Venn diagrams
* Seasonal calendars/daily routines/activity profiles
* Ranking/scoring matrices
* Use of local value criteria and preference ranking
* Oral- and ethno-histories and local stories/portraits
* Community problem brainstorming
* Proxy indicators

Conclusion: Components of Participatory Land Management

Land management for sustainable development is not about finding the right science and the new technology that will solve our problems. The "technology kit-bag" approach to agricultural development in developing countries has created as many problems as it has provided solutions; and it has left many disappointed people, bypassed by development, because of their inaccessibility to scientifically determined solutions and the inappropriateness of solutions to their particular circumstances. The new agenda of farmer participatory land management is about matching existing knowledge – formal and informal – with the vast array of potential combinations of environmental circumstances and socio-cultural and economic situations. People have for centuries been gaining livelihoods from difficult environments; they have been learning all the time; the products of that learning have been handed down in the form of custom and practice; the accumulated knowledge cannot possibly be reproduced today in modern scientific experiments because of the huge number of possible permutations of circumstances and the complexity of relevant variables which affect the outcome of agricultural production.

What are the key components of a farmer-participatory approach to sustainable land management? Adapted from the listing of Hinchcliffe et al. (1995), formulated for the particular case of participatory watershed development, we may highlight the following differences between participatory and conventional approaches:

• *Participatory*: local communities are fully and actively involved in the analysis of their land-management problems. *Conventional*: communities may be consulted about their views, but the analysis is done by professionals.
• *Participatory*: external support organisation is a facilitator of analysis and a catalyst for action. For sustainability, it may assist with creation of local institutions and user groups to manage aspects of the natural resources;

and encourage such institutions to develop their own procedures, rules, capital, and operating criteria in order to ensure continuance after external support withdraws. *Conventional*: donor support creates new, imposed, externally financed structures with little linkage to local community or to local government.

- *Participatory*: information dissemination is by farmer-to-farmer extension and informal networking; the object is to create greater self-reliance and closer collaboration and community dependency; extension agents act as facilitators. *Conventional*: information is transferred by extension agents through key informants, demonstration plots, field visits, and the media; the aim is to convince farmers of the importance and utility of the information.
- *Participatory*: flexibility and adaptation to local circumstances pervade any recommended technologies and selected crops; individual farmer needs and criteria for choice override technical specifications. *Conventional*: development of blueprint solutions and recommended practices by professional staff for local people.
- *Participatory*: emphasis on sustainability, equity, and access to improvements; not on short-term benefits. Benefits which occur without subsidies, inducements, and external assistance are favoured. *Conventional*: adoption criteria include technical efficiency, production maximisation, cost-benefit analysis.

Our challenge in land management for sustainable development is to harness the knowledge that is right for the times, the people, and the environment – a knowledge that has been constructed from a diverse and complex set of causalities, and one that has withstood a society which is placing increasing demands on natural resources. Some may conclude from this listing that science is usurped and that myth and local knowledge take its place. However, science has its mythology too, not least in conclusions based upon poor measurement, inadequate data, and simple mistakes (e.g. the case of soil erosion: Stocking 1995a). The role of science has to be different: as observer rather than manipulator; becoming more socially aware, and accepting interdisciplinary social science as an equal in deriving applicable conclusions; rejecting technical fixes and replacing them with tentative hypotheses and lists of optional strategies. Of course, such a role is less satisfying to those who wish to pontificate: therefore, the style of science needs to be different – less dictatorial, more provisional, greater willingness to learn from others, less inclined to force its own recommendations.

In such changes which are implied in a farmer-participatory approach to land management, there still exists a crucial role of science to describe, understand, and seek explanations for practices and views of land users. This is what PLEC (People, Land Management, and Environmental Change) is all about. The project is searching for ways farmers have learnt how to cope with environmental change and other pressures (such as conflict, economic forces, and demographic change), principally through utilising variety and diversity in the natural environment. Biodiversity, for example, is seen in many small-farm management strategies in the ways that people promote it in their own lands and conserve genetic pools to cover for an uncertain and unpredictable future. PLEC, we believe, highlights the necessary and radically different approach to science, by working with anthropology, the humanities, and socio-economics as well as interdisciplinary science to find a much more realistic understanding of the complexities of the real world in which we all operate.

Notes

1. As Häusler (1995) points out, different approaches to indigenous and scientific knowledge systems may not be so contradictory because of the knowledge itself, but because of the perspective of the people promoting the knowledge. This has led in recent years to the "actor-oriented" approach, which asserts that the processes by which the social actors interact, negotiate, and accommodate to each others' life-worlds lead either to reinforcement of existing types of knowledge or to the emergence of new forms – see Long and Long (1992) for a discussion of the distinctions between knowledge systems and how these may affect development discourses.
2. Numerous publications now give guidance on methods of participatory learning and action. See, for example, Pretty et al. (1995) and the *RRA Notes* published by the International Institute for Environment and Development, London, between 1988 and 1995. From Number 22, February 1995, the *RRA Notes* changed name to *PLA Notes* (Notes on Participatory Learning and Action) in recognition that many of us are unhappy with the notion of "rapid" and cannot see why the discussion should be confined to "rural."

REFERENCES

Apfel-Marglin, F. and S.A. Marglin. 1990. *Dominating Knowledge: Development, Culture and Resistance*. Oxford University Press, Oxford.

Atampugre, N. 1993. *Behind the Stone Lines*. Oxfam, Oxford.

Barraclough, S. 1995. Social dimensions of desertification: A review of key issues. In: D. Stiles (ed.), *Social Aspects of Sustainable Dryland Management*, 21–79. J. Wiley, Chichester, UK.

Behnke, R.H. and I. Scoones. 1992. *Rethinking Range Ecology: Implications for Rangeland Management in Africa*. The World Bank, Environment Working Paper No. 53. Washington, DC.

Brookfield, H.C. 1993. Environmental sustainability in Southeast Asia: UNU regional research projects. In: J.I. Uitto and M. Clüsener-Godt (eds.), *Environmentally Sound Socio-economic Development in the Humid Tropics: Perspectives from Asia and Africa*, 83–100. United Nations University, Tokyo.

Chambers, R. 1983. *Rural Development: Putting the Last First*. Longmans, London.

Chambers, R. 1993. *Challenging the Professions: Frontiers for Rural Development*. Intermediate Technology Publications, London.

Chambers, R., A. Pacey and L.A. Thrupp (eds.). 1989. *Farmer First: Farmer Innovation and Agricultural Research*. Intermediate Technology Publications, London.

Critchley, W. and K. Sievert. 1991. *Water Harvesting: A Manual for the Design and Construction of Water Harvesting Schemes for Plant Production*. Report AGL/MISC/17/91, Food and Agriculture Organization of the United Nations, Rome.

EcoSystems. 1982. Southeast Shinyanga Land Use Study, 3 volumes. Report for the Shinyanga Regional Integrated Development Programme, Tanzania. EcoSystems Ltd., Nairobi.

Häusler, S. 1995. Listening to the people: The use of indigenous knowledge to curb environmental degradation. In: D. Stiles (ed.), *Social Aspects of Sustainable Dryland Management*, 179–188. J. Wiley, Chichester, UK.

Hinchcliffe, F., I. Guijt, J. Pretty and P. Shah. 1995. *New Horizons: The Economic, Social and Environmental Impacts of Participatory Watershed Development*. Gatekeeper Series No. 50. Sustainable Agriculture Programme, International Institute for Environment and Development, London.

IFAD. 1992. *Soil and Water Conservation in Sub-Saharan Africa: Towards Sustainable Production by the Rural Poor*. Report prepared by the Centre for Development Cooperation Services, Free University, Amsterdam. International Fund for Agricultural Development, Rome.

Kerr, J. and N.K. Sanghi. 1992. *Indigenous Soil and Water Conservation in India's Semi-arid Tropics*. Gatekeeper Series No. 34, Sustainable Agriculture Programme, International Institute for Environment and Development, London.

Long, N. and A. Long (eds.). 1992. *Battlefields of Knowledge*. Routledge, London.

Lundgren, L., G. Taylor and A. Ingevall. 1993. *From Soil Conservation to Land Husbandry: Guidelines Based on SIDA's Experience*. Natural Resources Management Division, Swedish International Development Authority, Stockholm.

Palmer, J. 1992. The sloping agricultural land technology experience. In: W. Hiemstra, C. Reijntjes and E. van der Werf (eds.), *Let Farmers Judge: Experiences in Assessing the Sustainability of Agriculture*, 151–163. Intermediate Technology Publications, London.

Pretty, J.N. 1995. *Regenerating Agriculture: Policies and Practice for Sustainability and Self-reliance*. Earthscan, London.

Stocking, M.A. 1995a. *Six Myths About Soil Erosion*. Paper to Conference "Escaping Orthodoxy: Environmental Change Assessments, Local Natural Resource Management and Policy Processes in Africa," Institute of Development Studies, Sussex, 1994.

Stocking, M.A. 1995b. Soil erosion and land degradation. In: T. O'Riordan (ed.), *Environmental Science for Environmental Management*, 223–242. Longmans, Harlow.

Thomas, D.S.G. and R.J. Allison. 1993. *Landscape Sensitivity*. J. Wiley, Chichester, UK.

3

Women Farmers: Environmental Managers of the World

Janet Henshall Momsen

Introduction

There is a growing debate about gender and the environment which high-lights women's roles in the use and management of natural resources (Brai-dotti et al. 1994). This debate has stimulated much development analysis and created greater awareness of the activities of women farmers. But there are dangers in conceiving of women's roles in relation to the environment in a partial, narrow, or static way. Seeing women as isolated environmental actors, separate from men, with an innate understanding of *Nature* can be very misleading. Current development policy initiatives are often based on this essentialist assumption that women's relationship with the environment is special and, therefore, women are particularly interested in and capable of protection of the environment. Such a view enables policy makers to argue that projects aimed at sustaining the environment will also benefit women, and vice versa. This synergistic approach can be seen as creating both a trap and an opportunity.

At the level of rhetoric and debate, it is widely understood that women, in their productive and reproductive roles, have close links with the environment in many countries and that they are often among the first to be affected by resource degradation. However, policy makers do not always appreciate the diversity and complexity of the relationship between women and the environment, resulting in unexpected failures in development projects. For example, a tree-planting project in Ethiopia, using women as labour, was seen by the funding agency as both improving the environment by reducing soil erosion and also assisting women by providing employment and addi-tional firewood. Local women, on the other hand, saw the tree planting as increasing their burden of work without improving their lives because men controlled the land and the trees (Berhe 1994). Thus an understanding of both property rights and the complexity of gender divisions of labour is vital to an appreciation of the link between women and the environment.

Access to Resources

In most countries modernization has been accompanied by a decline in women's entitlements to land and common property resources. Women are usually very dependent on common property resources for water, firewood, compost for farmland and wild herbs, mushrooms, fruits and nuts, as it is usually their responsibility to ensure that the family is supplied with these goods (Swope 1995). When these commonly held resources become scarce and property rights are exerted because of a perceived market value, their control tends to be assumed by men, although women's role as the supplier to the family of these resources does not change (Agarwal 1989). The process of land reform has often led to land ownership in male hands, or when land is granted to a household it is registered in the man's name so that women lose their traditional rights to land. Where women do have legal rights to land ownership and inheritance, the plots of land they are able to control are generally the smallest, least accessible, and least fertile (Momsen 1988). Usufruct rights are often considered to be separate from land ownership, and furthermore the ownership of trees and land may be in different hands. If women plant trees on family land in order to find a new cash-earning product and accessible firewood, as in the case of the shea-butter tree in Uganda, men see this action as a declaration of land ownership and so uproot the trees (Rukaaka 1994). In other cases, men and women may use individual trees in different ways which are not always compatible. When resources begin to be scarce this incompatibility becomes a problem, as has occurred in the Dominican Republic, where women use palm fronds for making baskets and the fruit for pig food, while the male property owner is the only person allowed to cut the tree for timber (Rocheleau 1994).

One of the reasons for the decline in women's access to resources is that both land redistribution and subsidized agricultural inputs are in the hands of men who see women as dependents rather than individuals. Where the distribution of resources is used to reinforce the dominant position of the controller of these resources there is little incentive to offer them to women who in most societies have very little influence or status. In northern Nigeria when the distribution of government-subsidized chemical fertilizer was put in the hands of village chiefs rather than extension agents, women farmers were no longer able to obtain it (Ajakpo 1995). They had to buy fertilizer at the full market price, which few could afford. Because agricultural extension workers had been very successful in demonstrating to farmers that chemical fertilizer was vital for crop production, women who were unable to obtain this input left their fields uncultivated. Recent research in East and West Africa has shown that women can approach or equal men's productivity in sub-sistence farming even with a smaller resource base. In studies of farmers in Kenya and Burkina Faso, when statistically adjusted for resource differences, women farmers proved to be more productive than men (Blumberg 1995). In Kenya it was found that the influence of schooling on output was greater for women than for men for individuals with fewer than four years of education, and women benefited less than men from the predominantly male agricul-tural extension service (Moock 1976). A more gender aware agricultural policy might well bring very positive returns in many developing countries.

Loss of access to resources is particularly problematic for the one-quarter to one-third of rural women who are household heads, *de jure* or de facto, and the further 10 per cent who live in polygamous households. These women are often explicitly excluded from land reform and public-housing projects because they are seen as having too few adult workers and being too poor. The number of households in which women are the sole family support is increasing worldwide, especially in rural areas because of male out-migration. In an additional 25 per cent of households women provide more than half the total income. It is in these households economically dependent on women that the feminization of poverty is seen most clearly and pressure on the environment is most acute.

Women's Agricultural Work

Women work in agriculture in all parts of the world, especially in Africa (fig. 1). In many countries they are considered merely as unpaid family helpers and so are invisible in official statistics. Women's agricultural work covers production, processing, preparation, and preservation of foodstuffs and other farm products. They are also often responsible for marketing pro-duce from the farm. Extension workers assume that each family has a single male decision maker, that economic benefits are shared equally within the household, and that only products marketed through official channels need to be studied. Field research has shown that these normative views of family structure, income distribution, and farm output need to be reconsidered.

The historic neglect by agricultural research and extension of home gar-dens as significant production sites is a particular aspect of the commodity focus which has contributed to the underestimation of women's agricultural activities. Home gardens typically are seen as an extension of women's household duties and therefore outside the public sphere. These gardens are a haven of biodiversity, for they operate as a source of early maturing staples, a reserve for plant materials and seeds lest field crops fail, conservation sites for preferred or special varieties, testing grounds for new varieties, and security for stock or poultry needing special care. Many of the plants are grown in precise locations, such as round the edges of the plot, as for exam-ple pigeon pea, *Cajanus cajan*, in the West Indies, or in symbiotic combina-tions and so are neither particularly visible nor, in terms of proportionate volume, significant. Yet their value to household food security and income

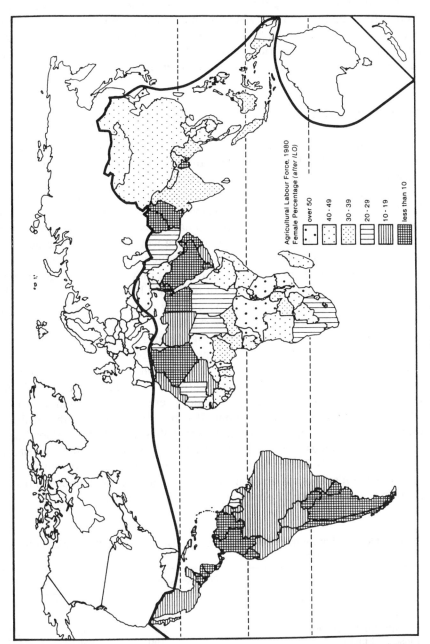

Agricultural Labour Force, 1980
Female Percentage (after ILO)

over 50

40 - 49

30 - 39

20 - 29

10 - 19

less than 10

Figure 1

may be high in terms of their functional or seasonal contribution, for example as a source of snack foods at times of year when women are too busy to cook frequently. These gardens, and the small livestock associated with them, often supply products which women trade locally and so they are an important source of women's cash income. As field crops become more and more cash- and export-oriented these home gardens become the major source of family subsistence, more so than official statistics indicate.

Time as a Resource

Time is a resource which is often very scarce for women farmers. Lack of time may mean that land is left uncultivated, farm operations are not performed at the optimal moment, and conservation measures not undertaken. Deterioration of the environment in terms of loss of available sources of firewood or pollution of water supplies will mean that women have to spend longer obtaining these resources and so have less time to spend on farming (Awumbila and Momsen 1995). In general, a woman's productive activities on the farm compete for time with her reproductive activities of bearing and raising children, managing the household, and playing a role in community management. Women in most parts of the world have a longer working day than men and so lack of time can be a major cause of declining food production. They are often used as a reserve supply of labour that can be called on at periods of peak demand in the agricultural calendar. At these times their work hours may reach 18 hours per day (table 1). This excessive work burden often occurs at periods of low food supplies and so leads to physiological stress. Women in many societies are last to eat at the family table and yet they have to combine the physiological demands of pregnancy and lactation with heavy agricultural labour. It is not surprising that the health of women in their reproductive years is worse than that of men. Poor health among women feeds into the cycle of poverty by constraining subsistence production. This is further limited by the traditional demand that the first call for women's labour is on men's fields, so that tasks on women's plots are either not done or carried out too late for optimal yields.

Table 1 Gender Division of Time Use in the Dry Zone of Sri Lanka (hrs./mo.) (Source: Wickramasinghe 1993: 170)

	Peak season		Slack season	
	Male	Female	Male	Female
Agricultural production	208	299	245	235
Household tasks	90	199	60	220
Fetching water and firewood	30	50	30	60
Social and religious duties	8	12	15	15
Total work hours	426	560	350	530
Leisure/sleep	294	160	370	190

Conclusion

There is no evidence that men farmers are any less aware of environmental problems than women. Gender studies of environmental perception in Thailand, Barbados, and China show that, when variations in age and education are held constant, problems such as soil erosion and deforestation are perceived in similar ways by men and women. Expecting women farmers to be the main managers of the environment not only overloads women with the responsibility of environmental protection but it also belittles the role of men. At the same time conflating women and the environment provides an excuse for aid agencies to reduce separate funding. There needs to be a better understanding of the flexibility and complexity of gender, of spatial variations in both environment and gender roles and of the time, labour, and financial pressures on poor households, predominantly female-headed households, which force farmers to ignore conservation requirements.

REFERENCES

Agarwal, Bina. 1989. Rural women, poverty and natural resources: Sustenance, sustainability and struggle for change. *Economic and Political Weekly* 28 October, pp. 46–65.

Ajakpo, Julie. 1995. *Gender and Agricultural Innovation in Northern Nigeria.* International Geographical Union Commission on Gender Working Paper No. 30. University of California, Davis.

Awumbila, M. and J.H. Momsen. 1995. Gender and the environment: Women's time use as a measure of environmental change. *Global Environmental Change: Human and Policy Dimensions* 5:4, 337–346.

Berhe, Tseghereda. 1994. Personal communication, May.

Blumberg, Rae L. Introduction: Engendering wealth and well-being in an era of economic transformation. In: R.L. Blumberg, Cathy A. Rakowski, Irene Tinker, and Michael Montéon (eds.), *EnGENDERing Wealth & Well-Being: Empowerment for Global Change*, 1–14. Westview Press, Boulder, CO.

Braidotti, Rosi, Ewa Charkiewiwcz, Sabie Hausler and Saskia Wieringa. 1994. *Women, the Environment and Sustainable Development: Towards a Theoretical Synthesis.* Zed Books in association with INSTRAW: London.

ILO (International Labour Office). 1986. *Economically Active Population Estimates, 1950–2025.* ILO, Geneva.

Momsen, Janet Henshall. 1988. Gender roles in Caribbean agricultural labour. In: M. Cross and G. Heuman (eds.), *Labour in the Caribbean*, 141–159. Macmillan, Basingstoke, UK.

Moock, Peter R. 1976. The efficiency of women as farm managers: Kenya. *American Journal of Agricultural Economics* 58 (December): 831–835.

Rocheleau, Diane E. 1994. Lecture at UC Davis, May.

Rukaaka, Kamerwa. 1994. Personal communication, May.

Swope, Lindsey H. 1995. *Factors Influencing Rates of Deforestation in Lijiang County, Yunnan Province, China: A Village Study.* Unpublished M.A. thesis, Department of Geography, University of California, Davis.

Wickramasinghe, A. 1993. Women's roles in rural Sri Lanka. In: J.H. Momsen and V. Kinnaird (eds.), *Different Places, Different Voices*, 159–175. Routledge, London.

4

Land-Use Change and Population in Papua New Guinea

Graham Sem

Introduction

Papua New Guinea (fig. 1) comprises the eastern half of the island of New Guinea. It is a geomorphologically diverse country in the South-West Pacific Ocean and contains four large provincial islands and over 600 smaller islands. The total land area of the country is 459,854 km² (Saunders 1993) with enormous social, cultural, and biophysical diversity. The country is located on the boundary between the northward moving Australian continental plate, and the north-west moving Pacific plate, which makes it one of the tectonically active areas in the world. The main islands are characterised by block-faulted, folded, and mountainous interiors. The highest peak is Mt. Wilhelm in the Simbu province, which rises to 4,510 metres above sea level. The deltaic flood plains provide the largest areas of lowlands especially along the south coast, where freshwater swamplands are common.

Terrestrial habitats range from extensive lowlands with rainforest, savanna, grassland, and freshwater swamps to upland montane rainforests and alpine grassland (table 1). The marine and aquatic environments appear equally diverse. Papua New Guinea's native flora comprise an estimated 15,000 to 20,000 species of vascular plants, including ca. 2,000 species of orchids, and more than 2,000 species of pteridophytes (Johns 1993).

Papua New Guinea culture is richly varied and people have lived in the lowlands for at least 40,000 years (Groube et al. 1986) and in the highlands for more than 24,000 years (White and O'Connell 1982). More than 750 different linguistic groups have been identified, with a variety of cultural responses to the environment. Great linguistic diversity in PNG is unmatched

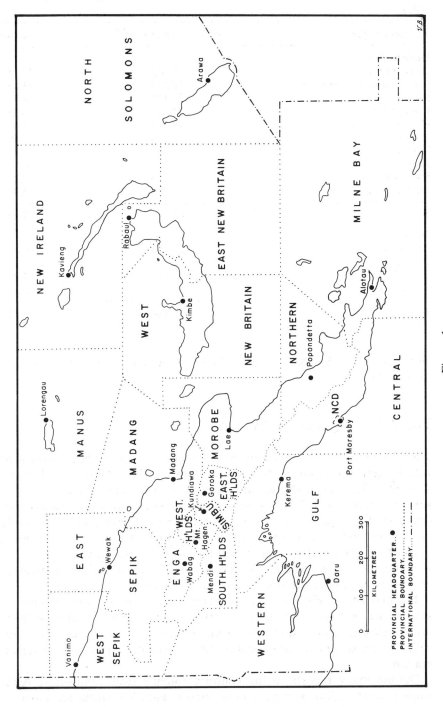

Figure 1

Table 1 Areas of Various Vegetation Types in Papua New Guinea (after Beehler 1985)

Vegetation type	Area of cover (km²)	% cover
Undisturbed		
Lowland rainforest	110,615	33.76
Lower montane rainforest	76,180	23.26
Savanna	22,120	6.76
Palm swamp	21,010	6.39
Herbaceous swamp	19,535	5.97
Mangrove	4,800	1.49
Montane forest	4,695	1.43
Alpine	1,720	0.53
Swamp forest	710	0.22
Strand forest	205	0.06
Total	**261,590**	**79.87**
Disturbed		
Grassland	27,180	8.30
Gardens	21,810	6.66
Degraded forest	15,910	4.80
Plantations	1,200	0.37
Total	**66,100**	**20.13**

elsewhere and it has been suggested that the number of languages used is likely to be over 800.

Traditional Agriculture System

The making of history in Papua New Guinea has always been associated with its agriculture. Over 250 food-plant species have been recorded and 43 of these have been always cultivated, 51 are cultivated and harvested as wild resources, and 157 are gathered from forests, savannas, and grasslands (Paijmans 1976). Most people are subsistence agriculturalists growing mainly tuber crops and planting some fruit and nut trees. The period of cropping ranges from three to five years in the lowlands to continuous cultivation in some highland areas. The fallow period ranges from no fallow to about 25 years' fallow, although in some areas the fallow period is up to 50 years.

The majority of PNG agriculture systems are present fallow systems, or systems which have evolved from forest fallow systems. Fallow systems involve clearing and cutting forest, some burning of felled vegetation, cultivation of crops, and abandonment of the site to natural processes of regeneration.

Despite many reports to the contrary (mainly from South-East Asia) tropical forest fallow systems can be stable. Fallow systems are environmentally friendly because a tree cover over often erodible and naturally poor soils is maintained. Stable shifting agriculture systems do not destroy rainforest by cutting and burning trees; they cycle through secondary forest. Shifting cultivators avoid using previously uncultivated forests, if possible, because of the difficulties and dangers of felling the trees, and because generally secondary forest contains more useful plants and animals than primary rainforest.

Latitudinal differences in solar energy, temperature, rainfall, and soil nutrients are some of the ecological constraints on agricultural systems (Bayliss-Smith and Feachem 1977). Often a farmer is unable to directly control the constraints of solar radiation, temperature, and rainfall but is able to adapt to changes brought about by these constraints. Papua New Guinea agriculturalists have been able to adapt to changes brought about by climate change and the socio-economic conditions so that agriculture has remained the mainstay of rural societies for over 3,000 years. Some of these responses and mitigating factors have resulted in the development of most elaborate farming techniques, such as mounding, terracing, mulching, ditching, draining, and irrigation systems which are now considered to have developed independently of the major agricultural areas of the world. Some writers ascribe development of such techniques to the introduction of new crops, population growth, and increased demands for social production (*sensu* Brookfield 1972) and a combination of all of these factors. This technological change/innovation was a result of agricultural intensification that is now evident in Papua New Guinea.

Land Use before Independence

Most land-use observations before independence in 1975 were those derived from other investigations such as gold exploration and copra-plantation establishment. However, more systematic investigations of the geology and landscape, as well as land evaluation investigations were carried out by the Commonwealth Scientific and Industrial Research Organisation (CSIRO), Division of Land Use Research. The methodology for land-evaluation investigations was based on the land systems survey (Christian and Stewart 1953). A land system is defined as a unique assemblage of the features of land, such as soils, vegetation, land-forms, rainfall, land use, and population. These surveys were designed mainly to assess the suitability of land for mechanised agriculture and pastoralism; and were based on air-photo interpretation and field checking which was limited to a representative selection of each area of the country investigated.

A total of 15 land systems were surveyed covering about 50 per cent of the total land area of the country. Land use was divided into three broad categories, subsistence cultivation, cash cropping (indigenous), and plantation. Most people in each land system were involved in subsistence cultivation. Cash cropping of mainly tree crops, such as coconuts, cocoa, coffee, and some rice, was practised by the indigenous population in both lowland

and highland areas, although in the latter some pyrethrum (*Pyrethrum cinerariifolium*), passion fruit, tea, and some livestock were introduced. Plantations were owned and managed by the non-indigenous population in both lowland and highland areas.

Based on land-system surveys undertaken between 1954 and 1976, it was estimated that 20 per cent of PNG land was under shifting cultivation. Some estimates of areas were given for different categories of land use in lowland and highland areas. For instance, area under subsistence cultivation per capita was estimated to be between 0.08 and 0.24 ha, with an average of 0.12 ha (McAlpine 1967, 1970). The total land area brought into commercial production (i.e. plantations) by non-indigenous people ranged from 6,400 ha in East Sepik for coconut and cocoa to over 148,000 ha for coffee and pastures in the highlands. These figures may have since changed due to the increase in population accompanied by greater economic and social demands.

Owing to the incompleteness of the population figures between 1964 and 1976, it was difficult to obtain figures for the whole country. Even if in all land-system surveys land-use intensity was not clear, it was defined for some areas, such as the highlands and the lower parts of the Sepik and Western provinces. Where sweet potato (*Ipomoea batatas*) was dominant in densely populated areas of the highlands, intensive, almost permanent short-fallow cultivation was evident. McAlpine (1970) suggested that the length of cultivation and fallow cycles differed greatly in the highland areas and was probably related to land pressure, environment, and cultivation techniques, although it was mentioned that no field measurements were made in support of this suggestion.

A number of major events during the pre-independence period saw an increase in the introduction of new crop varieties and emphasis on animal husbandry. The first major food and nutrition survey was conducted during 1947 and resulted in the introduction of improved pig and poultry strains and several major new projects were commenced. By 1951, the emphasis on food crops was shifted to plantation crops (McKillop 1976), but the distribution of seed of introduced vegetable crops continued.

Plantation agriculture commenced much earlier than the introduction of new food-crop varieties. The first legislation to encourage locals (Papuans) to grow cash crops was formulated in 1894 by the Lieutenant-Governor Sir William MacGregor almost 10 years after the British had annexed Papua. In 1903, indigenous cash cropping became the mainstay of the colonial government's policy concerning agricultural development. Regulations were introduced for compulsory planting of economic trees such as coconuts, rubber, and citrus trees (Waddell and Krinks 1968).

Prior to the Second World War, expatriate-owned plantations were a major source of cash-crop production, covering 24,705 ha of land. After the war, some of these plantations could not survive due to shortage of labour, low commodity prices, and high shipping costs (Crocombe 1964). Indigenous agriculture thus was to be the only hope for the future because it did not require hired labour and huge capital as was previously experienced on expatriate-owned plantations. Each village (or group of villages) was encouraged to plant cash crops as a group and, as new aspirations and increased demands for development became inevitable, the government then introduced a policy for increasing village production, which subsequently shifted away from group planting by encouraging each family to cultivate cash crops by the 1950s (Morawetz 1967; Waddell and Krinks 1968). Growth of interest in cash cropping among the people was stimulated also by the implementation of a variety of land-tenure and marketing schemes which were entirely directed towards indigenous producers. Thus, from 1951 onwards agricultural extension efforts concentrated exclusively on cash crops (Bourke et al. 1981). Land systems under various uses were therefore divided into subsistence cultivation, cash cropping, both indigenous and non-indigenous, and plantation systems which were exclusively expatriate-owned and have been described in the previous section.

As the push for cash-crop production increased so did the introduction and distribution of new food-crop varieties of sweet potato, cassava, and banana. Other crops such as peanuts, pineapples, mango, pawpaw, and guava became widely accepted in the villagers' diet (Bourke et al. 1981). During the 1970s government efforts to replace food imports were directed towards introduced vegetables, potatoes, rice and a greater emphasis on export cash-crop production. Vegetables such as tomatoes, cabbages, potato and sweet potato were cultivated on a large scale particularly for urban markets. Large-scale rice and sugar-cane production was being planned. Sugar production has been operated and managed by an overseas company since 1983.

Promoted by greater concern over increasing levels of malnutrition in rural PNG, more emphasis was placed on research and staffing, which resulted in a 1975 Nutrition Survey to assess, among other things, the levels of malnutrition in all provinces. This culminated in the formulation of the National Food and Nutrition Policy (NFNP). The main aim of the NFNP was to increase the proportion of total food supplies produced domestically. Much of the land was brought into some form of use because of the impending need for cash-crop and food production. Consequently, significant changes have occurred in food production during the last 40 years, which has led to other changes in food production and the nutrition system.

Population Dynamics

It has been difficult to understand the long-term population trends of Papua New Guinea mainly because of the fact that censuses have been conducted only for a relatively short period. Censuses were conducted in 1966, 1971, 1980, and 1990. Only the latter two attempted a complete enumeration of the rural and urban populations. The population of PNG has grown from 2.2 million in 1966 to an estimated 3.8 million in 1990. This increase represents a growth rate of 2.3 per cent per annum. The estimated mid-year population for 1994 would be about 4.1 million (Hayes 1993). It appears that the population growth rate declined during the 1970s and rose slightly during the 1980s. Thus, the growth rate declined from an annual average of 2.6 per cent during the period 1966–1971 to 2.1 per cent in 1971–1980, and rose again slightly to 2.3 per cent between 1980 and 1990. The current officially calculated growth rate is 2.3 per cent, based on an intercensal average (Hayes 1993).

The data as presented suggest that the population growth rate has changed little over a period of 25 years, but census-based estimates of vital rates indicate that both the birth rate and death rate have declined over the same period. However, given the current population growth rate of 2.3 per cent, and the increasing threat of logging and conversion of primary rainforest, it is highly likely that more land will be brought into use by the turn of the century. The present fallow systems, which rely on low population densities and large areas of undisturbed forest, will be shortened. Assuming little or no technological change, shorter fallows will cause forest and land degradation and environmental stress. Farming of degraded lands will not be sustainable in the longer term without innovations such as introduction of new crops, new technology, and soil-fertility maintenance techniques.

The spatial distribution of the population indicates that the southern and north-western coastal regions have low population densities (4 and 7 persons/km^2, respectively), while the islands and highland regions are more densely populated (10 and 22 persons km^2, respectively). The Western, Gulf, and West Sepik provinces remain sparsely populated. Over one-third of PNG's population is concentrated in the 13.5 per cent of the total land area in the highland region. Although 22 persons/km^2 is the average density for the 10 highland regions, it is reported that in some fertile highland valleys densities exceed 200 persons km^2 (Allen 1984), and it is in these areas that reports of "population pressure" on land have been most frequent.

Patterns of Land Use

Data are not available to classify land use and relate it to population density in PNG, although recent attempts have been made to classify land use by degrees of intensity (Saunders 1993). Agricultural land-use patterns vary greatly between provinces and within provinces where generalizations might misrepresent some areas (table 2).

The average cultivated area and land under use (20 per cent and 24 per cent, respectively) for the southern region is lower than the highland region (40 per cent and 43 per cent, respectively), while the north-western region has an average of 34 per cent cultivated and 42 per cent total land under use. The land area under some use in the island region is somewhat higher, with an average of 47 per cent cultivated land and 48 per cent land under some use. Thus, the total land area under any form of use is much greater than it is in the southern, north-western, and highland regions. The difference between

Table 2 Percentage of Total Land Area: Cultivated and Used Land by Province (Source: Saunders 1993)

Province	Total area (km²)	Population 1990	Persons per km² 1990	Total cultivated land (%)	Total used land (%)	% change
Western	97,065	110,420	1	8.0	10.0	2.0
Gulf	33,847	68,122	2	11.0	12.0	1.0
Central	29,954	141,241	5	21.0	30.0	9.0
Milne Bay	14,125	158,700	11	40.0	47.0	7.0
Oro	22,510	96,239	4	19.0	22.0	3.0
Southern Highlands	25,698	317,184	12	27.0	29.0	2.0
Enga	11,839	235,561	20	31.0	37.0	7.0
Western Highlands	8,897	333,828	38	50.0	55.0	5.0
Chimbu	6,022	186,114	31	42.0	43.0	1.0
Eastern Highlands	11,006	300,515	27	50.0	53.0	3.0
Morobe	33,525	377,563	11	36.0	43.0	7.0
Madang	28,732	256,370	9	56.0	64.0	8.0
East Sepik	43,720	255,012	6	20.0	37.0	17.0
West Sepik	36,010	140,051	4	23.0	26.0	3.0
Manus	2,098	32,840	16	83.0	84.0	1.0
New Ireland	9,615	86,999	9	47.0	47.0	0.5
East New Britain	15,109	185,024	12	25.0	25.0	—
West New Britain	20,753	130,625	6	27.0	31.0	4.0
North Solomons	9,329	155,000	17	55.0	55.0	—
Total	**459,854**	**3,567,408**		**25.0**	**30.0**	**5.0**

the total cultivated land and the total used land is less than 10 per cent (table 2 and fig. 2). However, the difference in the East Sepik province exceeds this figure (17 per cent change) and this difference is mainly attributed to sago (*Metroxylon sago*) collection from wild stands. Sago is an important staple food (starch) in the East Sepik and other parts of lowland PNG. Again the difference in the island region is minimal, except for the West New Britain province, where there is intensive logging, raising the difference between cultivated and used land to 4 per cent. Saunders (1993) has identified 12 land-use intensity classes and divided them into six cultivated and six un-cultivated land categories (fig. 3), which are described here. LU0 is defined as land with very high intensity tree crops (and others, such as sugar-cane) plantations. LU1 is also a very high intensity land use dominated by food-crop cultivation in areas of very high population density, permanent agri-culture, and cultivation cycles of over five years. LU2 is a high-intensity land use where food production is the primary use in high population-density areas, and semi-permanent cultivation in highland valleys and some parts of the lowland Madang and East Sepik provinces.

Land use of moderate intensity (LU3) occurs in areas devoted primarily to food production associated with moderate population density, with short to moderately long fallow periods. Low-intensity land use (LU4) is common in low population-density areas with moderately long fallow periods in all lowland provinces and some highland areas. Very low intensity (LU5 and LU6) of land use occurs in areas of low to very low population densities, especially in lowland areas where alternative sources of food (e.g. sago and fish) supplement the cultivated crops; and where gardens are widely scattered, especially in the Western province and parts of the West Sepik province.

The uncultivated land is classified as grassland (LU7), sago stands (LU8), subalpine grassland (LU10), and savanna woodland (LU11). Grasslands have been developed and maintained as a result of burning, while sago stands provide a staple food source in large wetland areas of East Sepik, Western, and Gulf provinces. Subalpine grasslands occur in inter-montane valleys and basins of the highlands between elevations of 2,500 and 3,000 metres above sea level, while alpine grasslands occur above the tree line (above 3,400 m). Savanna woodland areas are dominated by a distinct dry-season Aw (Köppen system) climate, mainly in the southern parts of the mainland and parts of the Markham valley.

Conclusions

PNG is an agricultural country and it is likely that its increasing population will continue to rely heavily on agriculture in the year 2000 and beyond. More land will be brought into production and it is in the dense population areas where population pressure is becoming a problem. Despite critical shortages of land that have been reported in the highlands, no systematic

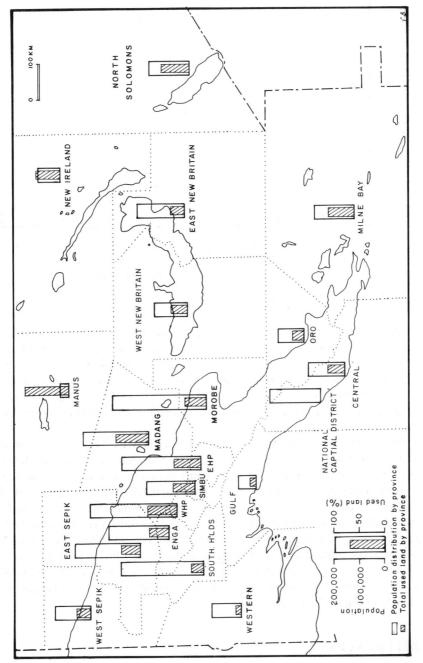

Figure 2

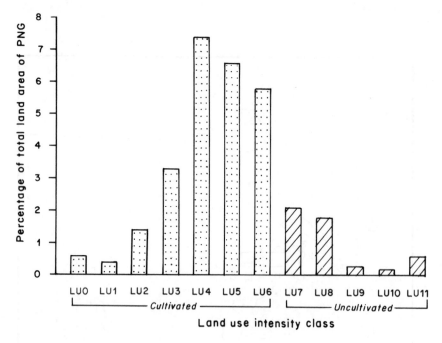

Figure 3

treatment has been undertaken to classify land use related to population density. However, recent data from the Agricultural Land Use Survey for PNG suggest that several areas could face similar problems of land use. Therefore, a greater research effort is required to assess and evaluate land-use change and to relate it to population growth in PNG. Such a research programme will ascertain the long-term sustainability of the current agricultural systems in the country.

At present, the Land Management Project (LMP) based at the Department of Human Geography, the Australian National University, is identifying, documenting, mapping, and describing agricultural systems for the whole country. Data on agricultural systems are currently being incorporated into a computer database linked to mapping software which, when completed for the whole country, will supplement biophysical data already contained in the Papua New Guinea Resource Information System (PNGRIS). The LMP will produce information at provincial and national levels on many aspects of subsistence agriculture, which will enable detailed assessment of the processes of intensification and the long-term sustainability of subsistence agriculture in Papua New Guinea.

REFERENCES

Allen, B.J. (ed.). 1984. *Agricultural and Nutritional Studies on the Nembi Plateau, Southern Highlands*. Department of Geography, University of Papua New Guinea.

Bayliss-Smith, T. and R.G. Feachem. 1977. *Subsistence and Survival: Rural Ecology in the Pacific*. Academic Press, London.

Beehler, B. 1985. *Conservation of New Guinea Rainforest Birds*. ICBP Technical Publication 4:233–247.

Bourke, R.M., B. Carrad and P. Heywood. 1981. *Papua New Guinea's Food Problems: Time for Action*. Department of Primary Industry Research Bulletin 29. Port Moresby.

Brookfield, H.C. 1972. Intensification and disintensification in Pacific agriculture: A theoretical approach. *Pacific Viewpoint* 13:30–48.

Christian, C.S. and A. Stewart. 1953. General Report on Survey of Katherine-Darwin Region 1946. *CSIRO Land Research Series Report* 1. Canberra.

Crocombe, R.G. 1964. *Communal Cash-cropping Among the Orokaiva*. New Guinea Research Bulletin 4. Australian National University, Canberra.

Groube, L., J. Chappell, J. Muke and D. Price. 1986. A 40,000 year occupation site at Huon Peninsula, Papua New Guinea. *Nature* 324:453–455.

Hayes, G.R. 1993. The Demographic Situation in Papua New Guinea and Its Policy Implications: An International Perspective. *Proceedings of the 19th Waigani Seminar*. University of Papua New Guinea. Port Moresby.

Johns, R.J. 1993. Biodiversity and conservation of the native flora of Papua New Guinea. In: B.M. Beehler (ed.), *Papua New Guinea Conservation Needs Assessment*, Volume 2, 15–76. Government of Papua New Guinea, Department of Environment and Conservation, Boroko, Papua New Guinea.

McAlpine, J.R. 1967. Population and land use of Bougainville and Buka islands. In: *Lands of Bougainville and Buka Islands, Papua New Guinea*. CSIRO Land Research Series No. 20. CSIRO. Canberra.

McAlpine, J.R. 1970. Population and land use in the Goroka-Mt Hagen Area. In: H.A. Hanntjens (ed.), *Lands of the Goroka-Mt Hagen Area, Territory of Papua New Guinea*. CSIRO Land Research Series No. 27. CSIRO. Canberra.

McKillop, B. 1976. *A History of Agricultural Extension in Papua New Guinea*. Department of Primary Industry Extension Bulletin 10. Port Moresby.

Morawetz, D. 1967. *Land Tenure Conversion in the Northern District of Papua*. New Guinea Research Bulletin 17. Australian National University, Canberra.

Paijmans, K. 1976. *New Guinea Vegetation*. Australian National University Press, Canberra.

Saunders, J. 1993. *Agricultural Land Use of Papua New Guinea: Explanatory Notes to Map*. PNGRIS Publication No. 1. AIDAB, Canberra.

Waddell, E. and P.A. Krinks. 1968. *The Organisation of Production and Distribution Among the Orokaiva*. New Guinea Research Bulletin 8. Australian National University, Canberra.

White, J.P. and J.F. O'Connell. 1982. *A Prehistory of Australia, New Guinea and Sahul*. Academic Press, Sydney.

5

Agricultural Sustainability and Food in Papua New Guinea

Ryutaro Ohtsuka

Introduction

Papua New Guinea is located between the equator and 12°S latitude and consists of the eastern half of the large island of New Guinea and the Bismarck Archipelago and many other small islands. The majority of the Papua New Guinea peoples have subsisted on cultivation of non-cereal crops, such as taro, yam, banana, and sweet potato, and/or exploitation of sago palm. One of the most striking human ecological characteristics in this country is seen in different population densities in association with the environments where they have lived and the major foods which they have grown and eaten. Moreover, the diversity of Papua New Guineans' adaptation has recently been accelerated by modernization influences, for instance acceptance and enlargement of cash-crop farming and increased rural–urban migration.

Traditional food-production systems in Papua New Guinea can be broadly classified into four categories: two in the high-altitude zone and two in the low-altitude zone (fig. 1). In the high-altitude zone, sweet potato–dependent agriculture has been dominant in the central Highlands and taro-dependent agriculture in the Highlands fringe. In contrast to the low-altitude zone, the high-altitude zone had long been free from malaria, the most life-threatening disease in this country (Riley 1983), and thus has been more densely populated. More important is the fact that compared with taro cultivators, the sweet-potato cultivators have kept higher population densities due to markedly high productivity of this crop (e.g. Golson 1977; Bayliss-Smith 1985) and their elaborate cultivation techniques (e.g. Waddell 1972). It is

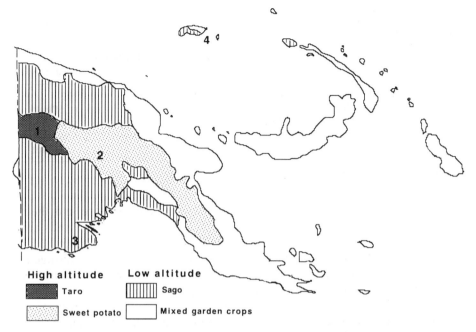

High altitude   Low altitude

▮▮▮ Taro   ▥▥ Sago

▒▒ Sweet potato   ☐ Mixed garden crops

Figure 1 The Geographical Distribution of Four Major Agricultural Types in Papua
New Guinea (Modified from King and Ranck 1982), and the Locations of the Four
Populations Treated in this Paper (1: Mountain Ok, 2: Huli, 3: Gidra, 4: Balopa)

also noted that acculturation has progressed more in the sweet potato–
cultivating area than in the taro-cultivating area (Feil 1987).

In the low-altitude zone, cultivation of various crops, such as taro, yam,
and banana, has been common in the eastern part of the main island and the
islands region, whereas exploitation of sago has been prevailing in the
western part of the main island. Sago making is characterised by high productivity per labour time (Ohtsuka 1985; Ulijaszek and Poraituk 1993) but
low productivity per total land area owing to the low densities of sago palms
which grow in a natural or semi-cultivated state. Consequently, the population density is lower in the sago-depending peoples than in other peoples
throughout the country (Townsend 1971; Townsend et al. 1982; Ohtsuka
1983). Another reason for the low population densities of the sago eaters, in
general, comes from delayed acceptance of acculturation influences.

The Sample Populations

The present paper first summarises the author's observations and findings in
four populations studied (fig. 1, table 1) and then discusses agricultural sustainability, taking into account demographic patterns and changes of the
people's lifestyle. The Mountain Ok villagers in the Highlands fringe, with

Table 1 The Four Selected Populations for the Intensive Study and Their Basic Characteristics

Name	Altitude	Modernization	Density (per km²)	Subsistence foods	Cash crops
1 Mountain Ok	High	Low	1.4	Taro	Nil
2 Huli	High	Medium	20–150	Sweet potato	(Coffee)[a]
3 Gidra	Low	Medium	0.5	Sago, tubers	(Rubber)
4 Balopa	Low	Low	30	Tubers	Cacao, peanut

a. Crops in parentheses are raised on a very small scale.

less acculturation influence, have carried out taro monoculture and their population density is 1.4 persons/km. The Huli in the central Highlands are one of the groups that developed elaborate sweet-potato cultivation techniques, for instance making mounds and/or ditches and planting nitrogen-fixing trees (*Casuarina spp.*) around the gardens. Their population densities are high, ranging from 20 to 150 persons/km² owing to different soil fertilities (Wood 1985). Because their land is far from the major urban centres, however, the Huli people have scarcely conducted cash-crop farming, contrasting to the bulk of the Highlands populations that have heavily been engaged in growing coffee for export and/or vegetables for marketing at the nearby local centres (Grossman 1984).

The Gidra form one of the typical sago-eating groups, though they have also grown tubers and bananas in the slash-and-burn gardens. Population density of the Gidra has been about 0.5–2 persons/km². They have had acculturation influences in the last several decades but their cash-earning activities have still been negligible. The author, together with his colleagues, has carried out repeated surveys among this group since the early 1970s, so that their long-term adaptation has been clarified (e.g. Ohtsuka and Suzuki 1990). The last group studied is the Balopa, who live in small islands in Manus Province. Their population density has been about 30 persons/km². The people, who have raised various garden crops for their subsistence use, began planting coconuts and cacao trees several decades ago, and recently grow peanuts for sale at the markets of Lorengau, the capital of Manus Province. Heavy out-migration is another characteristic of the Balopa people.

Agricultural Sustainability in Low Population Density Areas

The Gidra have depended for their plant food on sago starch and, to a lesser degree, garden crops. One of the 13 villages, called Wonie, was studied in detail. As shown in table 2, the Wonie villagers' energy intake from plant foods, which account for more than 90 per cent of the total intake, in three different periods, with source foods broken down in accordance with subsistence activities, can be characterised as follows. First, sago starch has

Table 2 Energy Intake (kcal) from Plant Foods per Day per Adult Male in Gidra-speaking Wonie Village in Three Time Periods

	1971	1981	1989
Wild plants	117	2	22
Sago	1,827	1,551	1,479
Garden crops			
Taro	439	556	0
Yam	16	352	772
Banana	314	76	293
Sweet potato	11	364	43
Others	138	243	214
Coconuts	317	10	332
Purchased	0	216	215
Total	**3,179**	**3,371**	**3,360**
(incl. animal foods)	(3,323)	(3,550)	(3,642)

provided almost half of the total energy, decreasing gradually throughout the periods; the decrease in dependence on sago is discussed later in relation to the recent population increase.

Second, in contrast to the fairly stable output from sago, that from various garden crops and coconuts changes over time. Taro had been the most consumed garden crop in this village for a long period, but it had completely vanished from the people's diet in 1989, as a result of the attack of the taro leaf blight (*Phytophthora colocasiae*), which has been widespread through-out Papua New Guinea (Purseglove 1972; Bourke 1982) and entered the Gidraland in the middle 1980s. The low banana consumption in 1981 reflected heavy damage to the crop by birds, which ate the immature fruit; this type of damage occurs every decade or so to different degrees. Similarly, because immature coconuts were almost completely eaten by rats (*Uromys caudimaculatus*) for several years from 1981, coconut consumption was negligible at the time. These damages were largely offset by the increased production of yams. On the other hand, the increased consumption of sweet potatoes in 1981 was derived from the introduction of high-yielding varieties in the preceding year through the local agricultural office. However, the villagers' cultivation techniques were not adequate to maintain productivity in the following years.

These year-to-year differences in food consumption, and thus in food production, in Wonie suggest two points relevant to agricultural sustainability, or stability of agricultural production. First, many local staples are not necessarily stable in productivity and are vulnerable to long-term natural environmental fluctuations. This implies, at the minimum, that mixed-crop or multi-crop farming contributes to avoiding deterioration in food production as a whole, as represented by the substitutional role of yams. Second, markedly high yields of sweet potatoes in 1981 (the yields in 1980 were much higher, though quantitative information was not obtained) suggest that the introduction of highly productive varieties or strains of food crops makes drastic increase of food production possible, if the appropriate agricultural technologies are associated.

As shown in table 2, the amount of sago consumed has decreased in Wonie. The author's estimation of their food consumption in the period without manifest acculturation influences, i.e. before the 1950s, reveals that sago consumption was greater (Ohtsuka 1993). The most plausible reason for the decrease of sago consumption and the increase of garden-crop consumption derives from the recent population increase. According to the author's genealogical–demographic analysis, the population increase rate per year markedly increased from only about 0.2 per cent before the 1950s to 2.0 per cent or over at present, a level similar to the national average in the 1970s or 1980s, owing to the provision of health services, immunization practices in particular. The changed population increase rate implies that the population has doubled in the period from 1960 to the present time (Ohtsuka 1986, 1990). To cope with such a high population increase, the Gidra people should have depended more on horticulture than sago exploitation because in their thinly populated land, the garden area can be easily expanded, whereas transplanting of sago palms, the bulk of which grow naturally, cannot meet with great success.

Gradual change of the staple food was also observed in the Mountain Ok villages, from taro to sweet potato (Kuchikura 1990; Ohtsuka 1994; cf. Bayliss-Smith 1985), as demonstrated in table 3, which compares percentage contributions of foods to energy and protein intake between a basically traditional taro-depending village and two villages which accepted sweet-potato cultivation. The cultivation of sweet potato is favourable to cope with increased population because the sweet potato is capable of producing higher yields owing to a short fallow period required and tolerating lower temperatures and poorer soils than taro. The gradual transitions from sago to garden crops among the Gidra and from taro to sweet potato among the Mountain Ok indicate that transfers of agricultural technologies, even those which were developed within Papua New Guinea, tend to elevate productivity in thinly populated and less-modernised societies.

Agricultural Sustainability in High Population Density Areas

In densely populated areas, agricultural productivity per land area, in general, has been high. Among the Huli, who have heavily depended on sweet-potato cultivation, their cultivable lands have already been exploited and the fallow period has been extremely reduced, eventually leading to permanent cultivation in many locations. As Wood (1985) suggested, soil erosion became manifest in the Huli land, even in the 1950s. Consequently, it seems difficult for them to maintain present agricultural productivity unless their agricultural system is modified.

Among the Balopa, whose major traditional crops for subsistence use are

Table 3 Percentage Energy Contributions of Five Categorised Foods in Three Mountain Ok Villages Observed in 1986

	Selbang[a]	Woktembip[b]	Fakobip[b]
Taro	56.1	13.1	3.6
Sweet potato	32.8	53.8	78.2
Other crops	9.2	30.3	10.1
Wild plants	0.2	2.7	8.1
Animal foods	1.7	0.1	1.5

a. Ohtsuka 1993
b. Kuchikura 1990

taro and yam and whose population density is about 30 persons/km^2, cash-crop farming was begun several decades ago. The early efforts were devoted to planting coconuts and then cacao trees. Because of the price reduction of these crops, however, they now grow peanuts, in particular, to sell at the local markets in the nearby town. Their livelihood has been assured by cash-earning agriculture, together with agriculture for subsistence use; in addition, the remittances from out-migrants, most of whom have high-level education and are engaged in qualified jobs. According to the author's observations, soil degradation has not seriously progressed because of the concentrated use of smaller areas for highly productive cash crops, in a sense that the productivity is assessed by not only nutrients contained in foods consumed but also money gained through the cash crops.

The situations of the two groups in the more modernised and densely populated areas suggest that it is possible for the people to survive based not exclusively on subsistence farming but on both subsistence and cash-crop farming. For maintaining earnings from cash cropping, economic development of the nation as a whole and well-planned agricultural policies are indispensable.

Population Increase and Agricultural Sustainability

Figure 2 shows a schematic diagram of changing population and carrying capacity in the four populations. In Papua New Guinea, the population increase rate was markedly elevated by the provision of health services, particularly immunization (Riley and Lehmann 1992). This change took place after the end of World War II, varying in the onset time from location to location but irrespective of the developmental degree of agricultural practices. For instance, among various societies such as the Gidra and the Mountain Ok, mortality rate was drastically improved during the time of low agricultural productivity (Ohtsuka 1993, 1994).

The original definition of carrying capacity does not include the role of cash earnings, but the author will use this term as the potential number of persons supported by agricultural products, utilised not only as food but also

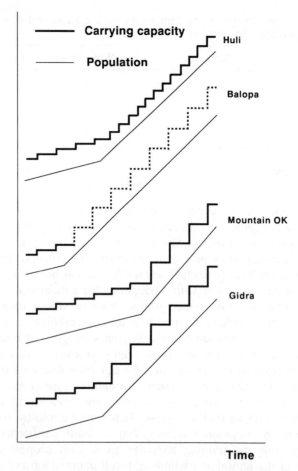

Figure 2 Schematic Diagrams of Changing Population and Carrying Capacity in Four Populations Studied. All Scales Are Arbitrary, though the Relative Inter-population Differences in the Population (or Population Density) and the Timing of Change to High Increase Rate Are Based on Empirical Information. The Dotted Line of Carrying Capacity for the Balopa People Reflects Both Subsistence Production and Earnings from Cash and Marketing

for cash earning, as represented by a dotted line; the contribution of cash earnings to carrying capacity was observed only in the Balopa among the four populations treated in this paper.

The scales of both population and carrying capacity are arbitrary, but the relative inter-group differences in population size and the onset time of change to a higher population increase rate are based on empirical information. Thus, it is noticed that the three populations other than the Balopa have not reached the point at which the cash economy plays a modifying role in carrying capacity.

As mentioned previously, the Gidra and the Mountain Ok peoples have gradually changed their agricultural systems to cope with population increase. Since their population densities are only 1 to 2 persons/km^2, their population numbers have still been lower than the carrying capacity. In other words, there is still room to raise their carrying capacity, mainly through technological changes. In contrast, serious problems took place in the densely populated areas. In particular, the Huli people have elevated their carrying capacity since long ago, mainly owing to the reduction of the fallow period of sweet-potato cultivation to attain high yields per land area in the long run. As a result, the soil fertility in their land has deteriorated. Thus, there is a high possibility that they cannot maintain the present productivity levels in the near future. The situation of the Huli is fairly contrary to that of the Balopa, who have begun cash cropping and whose land has not met with serious degradation.

As a conclusion, the author proposes solutions to cope with the accelerated population increase in the future. For less-modernised and thinly populated areas, the people should change or modify their agricultural systems to achieve more stable and higher productivity. For more modernised and densely populated areas, the people should extend cash-crop farming, simultaneously maintaining subsistence production. Two matters are essential in these processes. One is the introduction of high-yielding crops and elaborate technologies. The other is the enlargement of mixed-crop and/or multi-crop farming. These may be accomplished through the efforts of the local peoples themselves, in addition to the efforts of the governmental and non-governmental professionals. Also needed are national-level efforts to harmonise the development of the rural and urban sectors. Finally, the current population increase rate (2.7 per cent per year in 1990, according to the Population Reference Bureau) should be reduced as shortly as possible; otherwise, any efforts for agricultural development cannot produce fruitful results.

REFERENCES

Bayliss-Smith, T. 1985. Subsistence agriculture and nutrition in the Bimin Valley, Oksapmin Sub-District, Papua New Guinea. *Singapore Journal of Tropical Geography* 6:101–105.

Bourke, R.M. 1982. Root crops in Papua New Guinea. In: R.M. Bourke and V. Kasavan (eds.), *Proceedings of Second Papua New Guinea Food Crops Conference*, 51–63. Department of Primary Industry, Port Moresby.

Feil, D.K. 1987. *The Evolution of Highland Papua New Guinea Societies*. Cambridge University Press, Cambridge.

Golson, J. 1977. No room at the top: Agricultural intensification in the New Guinea Highlands. In: J. Allen, J. Golson and R. Jones (eds.), *Sunda and Sahul: Prehistoric Studies in Southeast Asia, Melanesia and Australia*, 601–638. Academic Press, London.

Grossman, L.S. 1984. *Peasants, Subsistence Ecology and Development in the Highlands of Papua New Guinea*. Princeton University Press, Princeton.

King, D. and S. Ranck. 1982. *Papua New Guinea Atlas*. Robert Brown, Bathurst.

Kuchikura, Y. 1990. Subsistence activities, food use, and nutrition among the Mountain Ok in central New Guinea. *Man and Culture in Oceania* 6:113–137.

Ohtsuka, R. 1983. *Oriomo Papuans: Ecology of Sago-Eaters in Lowland Papua.* University of Tokyo Press, Tokyo.

Ohtsuka, R. 1985. The Oriomo Papuans: Gathering versus horticulture in an ecological context. In: V.N. Misra and P. Bellwood (eds.), *Recent Advances in Indo-Pacific Prehistory*, 343–348. Oxford and IBH, New Delhi.

Ohtsuka, R. 1986. Low rate of population increase of the Gidra Papuans in the past: A genealogical-demographic analysis. *American Journal of Physical Anthropology* 71:13–23.

Ohtsuka, R. 1990. Fertility and mortality in transition. In: R. Ohtsuka and T. Suzuki (eds.), *Population Ecology of Human Survival: Bioecological Studies of the Gidra in Papua New Guinea*, 211–218. University of Tokyo Press, Tokyo.

Ohtsuka, R. 1993. Changing food and nutrition of the Gidra in lowland Papua New Guinea. In: C.M. Hladik, A. Hladik, O.F. Linares, H. Pagezy, A. Semple and M. Hadley (eds.), *Tropical Forests, People and Food*, 257–269. UNESCO, Paris.

Ohtsuka, R. 1994. Subsistence ecology and carrying capacity in two Papua New Guinea populations. *Biosocial Science* 26:395–407.

Ohtsuka, R. and T. Suzuki. 1990. *Population Ecology of Human Survival: Bio-ecological Studies of the Gidra in Papua New Guinea.* University of Tokyo Press, Tokyo.

Purseglove, J.W. 1972. *Tropical Crops Monocotyledons.* Longman, London.

Riley, I.D. 1983. Population change and distribution in Papua New Guinea: An epidemiological approach. *Journal of Human Evolution* 12:125–132.

Riley, I.D. and D. Lehmann. 1992. The demography of Papua New Guinea: Migration, fertility, and mortality patterns. In: R.D. Attenborough and M.P. Alpers (eds.), *Human Biology in Papua New Guinea: The Small Cosmos*, 67–92. Oxford University Press, Oxford.

Townsend, P.K. 1971. New Guinea sago gatherers: A study of demography in relation to subsistence. *Ecology of Food and Nutrition* 1:19–24.

Townsend, P.K., L. Morauta and S.I. Ulijaszek. 1982. *Sago Research in Papua New Guinea.* IASER Discussion Paper No. 44, Institute of Applied Social and Economic Research, Port Moresby.

Ulijaszek, S.J. and S.P. Poraituk. 1993. Making sago in Papua New Guinea: Is it worth the effort? In: C.M. Hladik, A. Hladik, O.F. Linares, H. Pagezy, A. Semple and M. Hadley (eds.), *Tropical Forests, People and Food*, 271–280. UNESCO, Paris.

Waddell, E. 1972. *The Mound Builders: Agricultural Practices, Environment, and Society in the Central Highlands of New Guinea.* University of Washington Press, Seattle.

Wood, A.W. 1985. *Stability and Performance of Huli Agriculture.* Department of Geography Occasional Paper No. 5. University of Papua New Guinea, Port Moresby.

6

Population Pressure and Agrodiversity in Marginal Areas of Northern Thailand

Kanok Rerkasem

Introduction

The highlands are among the most marginal areas in northern Thailand. Development has come in the past 30 years mainly due to national and international concerns on illicit opium cultivation by the ethnic minorities, and a national campaign against communist insurgencies. Many changes have taken place in the highlands in the last 30 years, some due to development efforts, others spontaneously. Other causes and consequences of highland development, however, are less clearly understood. A better understanding of current changes is essential for the future of sustainable development of the area.

One of the major forces behind changes in the highlands is population increase. High rates of population increase can be seen extensively in most of the highland villages. The pressure on the land is such that shifting agriculture is no longer feasible in many parts of the highlands.

Preliminary studies have suggested that this internal pressure of population increase combined with external pressures and opportunities has led to the move away from the traditional land use of shifting cultivation to more permanent land use with various levels of cropping intensification, for example multiple cropping with irrigation, mixed annual cropping of cash crops with soil and water conservation measures or fruit trees, and agroforestry systems. The advancement of the alternatives to traditional shifting cultivation has been largely depending on the local people, who have to simultaneously adapt their traditional farming systems to meet market opportunities as well as developmental policy changes. Inevitably, certain farming practices are "successful" while others have failed. Some successes

55

are private, with a high public cost in the form of environmental and watershed degradation. Others are economically successful, while also contributing towards forest and watershed conservation.

Within the framework of the United Nations University's collaborative research on People, Land Management, and Environmental Change (PLEC), the development of production pressure and agrodiversity paradigms will be examined at the village level. Preliminary results from selected villages with productive and sustainable strategies are presented.

Background

Until relatively recently, the concentration of the population in northern Thailand had been mostly in the alluvial plains of its major rivers (fig. 1). Although they account for three-quarters of the total land area, the highlands were largely sparsely populated. The population of the highlands has always been made up largely of people who belong to one of the ethnic minority groups (table 1). Among the major groups are Karen, Hmong, Lahu, Akha, Lisu, and Yao. These peoples form part of a much larger population of similar ethnic background that populates the mountainous area that covers the south-western provinces of China, Laos, northern Viet Nam, Myanmar, and Thailand (fig. 2). Collectively this area could be called "Montane Mainland South-East Asia" (MMSEA). The people of this area have made their living largely from agriculture, traditionally in the form of shifting agriculture. Land use in the highlands, however, has undergone major changes, especially in the past 30 years.

In Thailand, highland development policy in the 1960s through to the early years of the 1980s was driven by two major concerns:
1. national and international concerns over illicit opium cultivation by the ethnic minorities, and
2. a national campaign against communist insurgencies.

These have translated into programmes that have resulted in:
• social integration, i.e. granting of Thai citizenship and the beginning of

Table 1 Cultural Diversity of the Ethnic Minorities in Northern Thailand (Source: Matisoff 1983)

1. Sino-Tibetan (Miao/Yao)	2.	Tibeto-Burman
Hmong	2.1	Loloish
Yao		Lahu
Lua		Lisu
H'tin	2.2	Southern Lolo
Khmu		Akha
	2.3	Lolo-Burmese
		Karen?

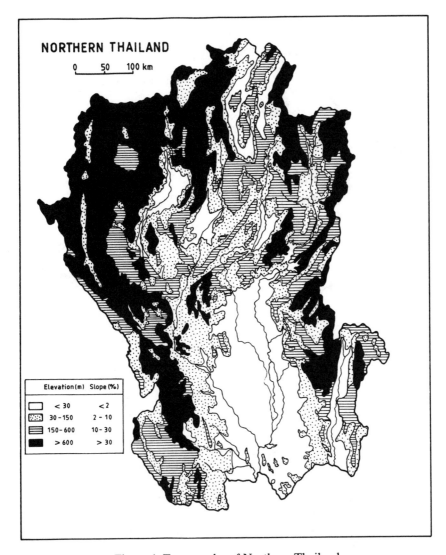

Figure 1 Topography of Northern Thailand

the process to integrate highland villages into the regular administrative structure of government through the hierarchy of village, subdistrict, district, and provincial office of the Department of Local Administration, Ministry of Interior;
- improving road access and electricity supply;
- relocation of a number of villages away from the influence of the insurgence; and
- opium substitution (with alternative cash crops) and eradication (by strict

Figure 2 Distribution of Lowland Rice and Upland Areas in South-East Asia. Shaded Area Represents the Major Area of Montane Mainland South-East Asia (MMSEA) (Source: Adapted from Garrity and Sajise 1993)

law enforcement involving the army in opium poppy crops destroying operations).

At the same time, another national policy to conserve national forests and watersheds was implemented by the Royal Forestry Department. This has resulted in the establishment of national parks and wildlife reserves (totalling almost two million hectares), and designation of areas to be strictly conserved as watersheds. The major effect this has on highland land use is the restriction of access to forest land by the highland's agricultural population.

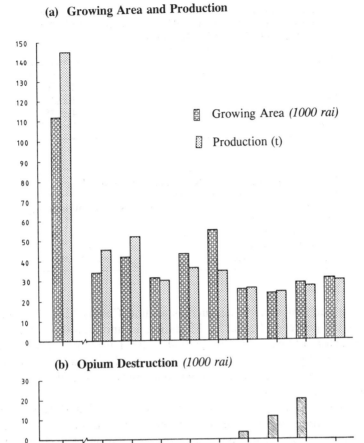

(a) Growing Area and Production

Growing Area *(1000 rai)*

Production (t)

(b) Opium Destruction *(1000 rai)*

1965/66   80/81 81/82   82/83 83/84 84/85 85/86   86/87   87/88   88/89

Figure 3 Changes in (a) Growing Area and Production of Poppy Opium and (b) Crop Destruction, 1965–88 (Sources: ONCB 1980 and 1988; ONCB/UNFDAC 1983)

1 *rai* = 0.16 ha

Dynamics of Highland Land Use

Many changes have taken place in the highlands in the last 30 years, some due to development efforts, others spontaneously. Illicit opium production has generally been satisfactorily curtailed through a combined effect of alternative cash cropping and strict law enforcement (fig. 3).

In addition to the "external" factors responsible for changes in the highlands mentioned above, another force was operating from within the highlands. This is the pressure exerted on the land by population increase (table 2). The population of the highlands in northern Thailand is now approaching one million. Annual growth rate between 1986 and 1993 was estimated

Table 2 Highland Population in Northern Thailand (Sources: Kunstadter et al. 1978; Kunstadter 1983; NSO 1985, 1986, 1987; NSC/NESDB 1993)

Year	Population
1972	272,568
1973–77	306,391
1985–87	495,353
1991	749,353

Table 3 Population Growth in Northern Thailand

Category	Increase per annum (%)
Total population	1.1
Highland population	6.0 (1986–91)
Chiang Mai province	12.0 (1985–91)

to be 6 per cent. In some areas such as around Chiang Mai, the rate is an incredible 12 per cent (table 3). Migration is undoubtedly a major contributing factor, adding to an already high natural growth of about three per cent per year. Migration comes both from across the border from Myanmar, Laos, and sometimes as far as China, but upwards migration of lowland Thais is also significant (fig. 4).

Traditionally, shifting agriculture in northern Thailand had derived a degree of sustainability from the use of long fallow periods of 8–10 years to restore soil fertility and to keep down the population of pests and weeds. This land-use practice, however, requires much land. For example, a detailed study of a couple of Lua and Karen villages which practise sustainable rotational shifting agriculture indicated that a total of 5–6 hectares of land was required per head of the population. Pioneer shifting agriculture villages, largely belonging to ethnic groups of Hmong, Lahu, Yao, Lisu, and Akha, appeared to perceive the need to adapt to cope with the mounting pressure on the land even 30 years ago. Many of these began to buy and/or develop paddy land for wet rice and settle down to become sedentary already in the 1960s. Rotational shifting agriculture villages were already sedentary. Figure 5 presents the process in a schematic form and figure 6 shows the transect of land use in Ban Tuan of Mae Chaem district in Chiang Mai province. In the past, population increases had been managed by a group of households splitting off to form new villages. The pressure on the land is indicated by difficulties in finding new sites for these "daughter" villages. The government's forests and watershed conservation policy has further aggravated this increasing pressure on the land.

Under this increasing pressure on the land, three types of coping strategies have been identified:

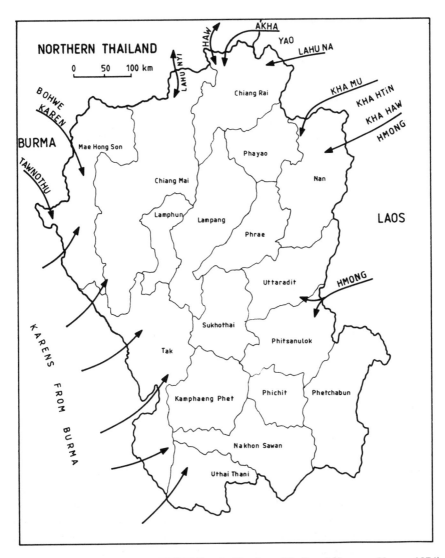

Figure 4 Migration Routes of Hill Tribes in Northern Thailand (Source: Young 1974)

1. People become economically poorer, their land and nearby forest become degraded.
2. Cash cropping is adopted, sometimes with economic success, sometimes with economic failure. The resource base becomes degraded and there may also be a negative impact on the forests and watershed.
3. Adoption of sustainable land-use alternatives that are economically viable, and also have positive impacts on the forests and watersheds.

The remaining part of this paper will focus on this third strategy, examining the mechanisms by which highland land use might adapt to various

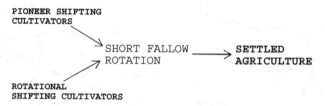

CURRENT STATUS OF SHIFTING CULTIVATORS

Figure 5 Changes from Traditional Shifting Cultivation to Permanent Agricultural Systems in Northern Thailand

external and internal pressures. Our preliminary studies have indicated that biological diversity plays a primary role in the development of sustainable land-use alternatives in the highlands.

Role of Biodiversity in Sustainable Use of Marginal Land

Two elements of diversity have been identified as central to the development of sustainable land-use alternatives.

Diversity of Livelihood Activities

A diversity of livelihood activities is a common feature of land use in the mountains of the MMSEA. Table 4 illustrates the degree of diversity in livelihood activities found in seven villages in northern Thailand. A "typical" farming system is engaged in six to ten activities. The performance and impact of a mountain farming system on the resource base and the environment is a combined effect of these numerous activities. Evaluating the performance of mountain farming systems and their impact may not be possible by looking at each individual activity, but has to take into account interactions among the different activities.

Agrodiversity and MMSEA PLEC in Northern Thailand

The primary hypothesis of the MMSEA PLEC project in northern Thailand is as follows:

In each farm household and each village in the highlands, these several livelihood activities are not carried out in isolation, but could be closely related. Synergies among different activities can be identified at different hierarchical levels of the agro-ecosystems, e.g. in a field, a farm, a whole village, and sometimes even among a number of neighbouring villages. (Rerkasem and Rerkasem 1995)

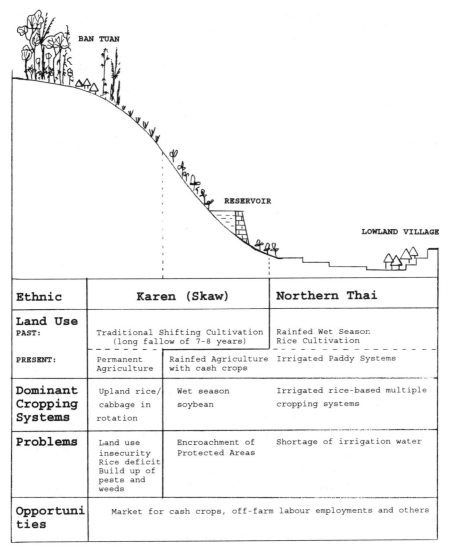

Ethnic	Karen (Skaw)		Northern Thai
Land Use PAST:	Traditional Shifting Cultivation (long fallow of 7-8 years)		Rainfed Wet Season Rice Cultivation
PRESENT:	Permanent Agriculture	Rainfed Agriculture with cash crops	Irrigated Paddy Systems
Dominant Cropping Systems	Upland rice/ cabbage in rotation	Wet season soybean	Irrigated rice-based multiple cropping systems
Problems	Land use insecurity Rice deficit Build up of pests and weeds	Encroachment of Protected Areas	Shortage of irrigation water
Opportunities	Market for cash crops, off-farm labour employments and others		

Figure 6 Transect of Land-use Change in Ban Tuan of Mae Chaem District, Chiang Mai Province (Source: CARE/HRIP 1994)

Examples of such interactions so far identified include for example gathering bamboo shoots from wild stands for sale in the Hmong village of Pah Poo Chom, north of Chiang Mai, which is kept below the level of "maximum sustaining yield" and is, thus, sustainable. This is partly because of a competing demand for labour for the cabbage crop (fig. 6).

Cabbage production was adopted in the Karen village at Mae Rid Pagae, in Mae Hong Son province of Thailand, as an alternative income-generating

Table 4 A Summary of Livelihood Activities in Seven Mountain Villages (Ethnic Groups: Hmong, Karen, Lahu, Lua) in Northern Thailand

Activity	Degree of involvement		Contribution to income (%)
	Number of villages	Households (%)	
Upland rice	7	76	10–80
Livestock	7	81	5–50
Wage earning	5	44	5–70
Wetland rice	4	64	25–70
Cabbages	3	52	5–50
Fruit trees	3	48	10–70
Soya beans	1	75	15–25
Chillies	1	95	10–50
Flowers	1	5	10
Minor crops	2	45	5–30
Teak harvesting	1	nd	nd
Opium	1	nd	nd
Swidden crops	5	100	nd
Gathering			
for subsistence	7	100	nd
for sale	3	60	5–30
Handicrafts			
for own use	7	100	nd
for sale	3	53	5–25

1. Data in table 4 were obtained from interviews with a sample of farmers, individually and in groups, in seven mountain villages in northern Thailand, representing 10–20 percent of the population in each village. The farmers were asked to list all daily activities considered to contribute to their living, and to give each a rating in contribution to their annual income, in cash and kind, relative to the most common crop, usually rice. Some activities were not quantified, i.e. not determined (nd). Household activities, like cooking, cleaning, washing, raising children, house repair, were not considered.
2. Number of villages in which the activity was found.
3. Percentage of households involved in the villages where the activity was practised.

activity. It has a further impact on land productivity, as the yield of rice that followed the cabbages was doubled or tripled, presumably as the result of residual effects of cabbage fertilizers. Intensification and commercialization of highland agriculture has effectively reduced pressure on the land, simply by increasing yield per area.

In Pah Poo Chom, which was on the brink of collapse in 1970 as the result of a decline in land productivity, cash cropping with irrigated cabbages and lychee combined with bamboo shoots gathering, has significantly alleviated pressure on the land and enabled the village to allow the forest to regenerate in the major part of its former shifting agriculture area. On the other hand, in many cases intensification and commercialization have also led to a loss of genetic resources of many domesticated species, for example sesame, chillies, legumes, and traditional vegetables, and indigenous knowledge on

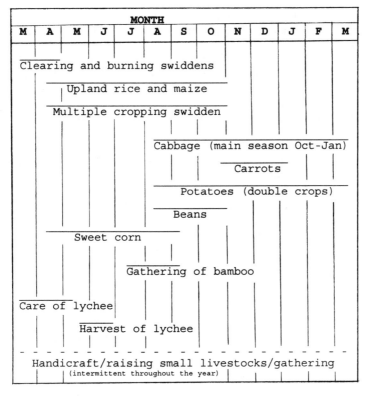

Figure 7 Activity Calendar of Pah Poo Chom Village (Source: Rerkasem and Rerkasem 1994)

their use and management, with a more direct effect on the loss of nutritional diversity and security.

Within the framework of PLEC, this hypothesis will be tested in more quantitative detail.

REFERENCES

CARE/HRIP. 1994. *CARE Highland Resources Integration Project, Phase III Implementation 1992–94. Final Evaluation Report.* CARE/International, Thailand, Bangkok.

Cooper, R.G. 1984. *Resource Scarcity and the Hmong Response: A Study of Resettlement and Economy in Northern Thailand.* Singapore University Press, Singapore.

Garrity, D. and P. Sajise. 1993. Sustainable land use systems research in Southeast Asia: A regional assessment. In: *Proceedings of a Workshop on Sustainable Land Use Systems Research*, 59–72. Rodale Institute, Kutztown, Penn., USA.

Guo Huijun. 1993. *Land Transformation: Farmers Adaptations in Southern Yunnan, China.* Paper presented at the PLEC Thailand/Yunnan Second Cluster Meeting, 4–6 November 1993, Kunming, China.

KIB. 1992. *Indigenous Landuse Management Systems in Yunnan*. Papers presented at a seminar on Traditional Land Management, 8–14 October 1992, Kunming Institute of Botany.

Kunstadter, P. 1978. Subsistence agricultural economics of Lual and Karen Hill Farmers, Mae Sariang District, Northwestern Thailand. In: P. Kunstadter, E.C. Chapman and S. Sabhasri (eds.), *Farmers in the Forest*, 71–133. The University Press of Hawaii, Honolulu.

Kunstadter, P. 1983. Highland populations of northern Thailand. In: J. McKinnon and W. Bhruksasri (eds.), *Highlanders of Northern Thailand*, 15–45. Oxford University Press, Kuala Lumpur.

Kunstadter, P., E.C. Chapman and S. Sabhasri (eds.). 1978. *Farmers in the Forest*. The University Press of Hawaii, Honolulu.

Matisoff, J.A. 1983. Linguistic diversity and language contact. In: J. McKinnon and W. Bhruksasri (eds.), *Highlanders of Northern Thailand*, 56–86. Oxford University Press, Kuala Lumpur.

Nakano, K. 1978. An ecological study of swidden agriculture at a village in northern Thailand. *Tonan Ajia Kenkyu* (South-East Asian Studies) 16:411–446.

NSC (National Security Council)/NESDB (National Economic and Social Development Board). 1993. A Directory of Highland Communities and Population, 1991. NSC and NESDB, Bangkok.

NSO (National Statistics Office). 1985. Survey of Hilltribe Population, 1985 Tak Province. NSO, Bangkok.

NSO. 1986. Survey of Hilltribe Population, 1985 Chiang Mai and Phayao Provinces. NSO, Bangkok.

NSO. 1987. Survey of Hilltribe Population, 1985 Lam Pang, Sukhothai, Phetchabun, Phrae and Nan Provinces. NSO, Bangkok.

ONCB (Office of Narcotics Control Board). 1980. Opium Poppy Survey. ONCB, Bangkok.

ONCB. 1988. Opium Poppy Survey. ONCB, Bangkok.

ONCB and UNFDAC (United Nations Fund for Drug Abuse Control). 1983. A Masterplan for Opium Poppy Cultivating Regions of Thailand. ONCB and UNFDAC, Bangkok.

Rerkasem, K. and B. Rerkasem. 1994. *Shifting Cultivation in Thailand: Its Current Situation and Dynamics in the Context of Highland Development*. IIED Forestry and Land Use Series No. 4. International Institute for Environment and Development, London.

Rerkasem, K. and B. Rerkasem. 1995. Montane Mainland South-East Asia: Agro-ecosystems in transition. *Global Environmental Change: Human and Policy Dimensions*, Vol. 5, No. 4, 313–322.

Sam, D.D. 1994. *Shifting Cultivation in Vietnam*. Country Report, IIED Forestry and Land Use Series No. 3. International Institute for Environment and Development, London.

Souvanthong, P., S. Symmavong and K. Keosacksit. 1994. *Shifting Agriculture in Laos*. Paper presented at the Workshop on Shifting Agriculture, Chiang Mai University.

TDRI. 1994. *Assessment of Sustainable Highland Agricultural Systems*. Natural Resources and Environment Program, Thailand Development Research Institute.

Young, G. 1974. *The Hill Tribes of Northern Thailand: The Origins and Habitats of the Tribes Together with Significant Changes in Their Social, Cultural and Economic Patterns*. The Siam Society, Bangkok.

7

Managing the Resources of the Amazonian Várzea

Christine Padoch

Background

The floodplains of the Amazon river and its major white-water tributaries, the *várzea*, occupy a mere two per cent of the total area of the Amazon basin, but play a role far out of proportion to their relatively small size. Várzeas are often considered the most economically important of all Amazonian lands because of their good soils, their access to transport, and their relatively concentrated human populations. For centuries the floodplains have been very important areas of resource extraction, agriculture, and human settlement. Research indicates that many areas of the várzea were once densely settled by highly organised indigenous peoples. Most of those societies were destroyed in the early years of the European conquest of Amazonia. In contemporary Amazonia, however, most people, both rural and urban, continue to live in or close to the várzea.

The alluvial soils of the várzea, replenished every year by floods, are highly fertile. However, the same annual floods that deposit those rich sediments also make these lands difficult to exploit by modern agricultural methods. Few existing technologies deal adequately with the risks and opportunities the diverse and changeable floodplains present. Today water buffalo ranches increasingly occupy much of the floodplain's fertile lands, particularly along its lower reaches. This extensive land use together with highly intensive and heavily fertilized vegetable production on the outskirts of a few major cities are two conspicuous modern uses. Each has an environmentally and socially questionable present, and an uncertain future. Neither is economically feasible for the great majority of Amazonian farmers.

In areas like the várzea that are potentially very productive but are unsuitable for exploitation using modern agricultural technologies, it may be most important to look at traditional, smallholder production systems. Attention in Amazon research has turned increasingly to locally developed practices. Researchers have pointed out that indigenous peoples have, over centuries, worked out technologies to exploit and yet conserve resources, including tropical flora, fauna, and soils, and that some aspects of these technologies can be successfully adapted to modern needs and demands. In the Amazon floodplains indigenous peasant and tribal populations use a great variety of agricultural and agroforestry practices that effectively exploit the diverse resources and cropping opportunities the area offers.

In this paper I discuss three examples of locally designed várzea resource management that are much more than fascinating subjects for scientific study. Each of these systems incorporates much knowledge and experience that is useful in designing appropriate agriculture, agroforestry, and forest management efforts in Amazonia.

To find traditional resource-management patterns and to find the people who best know the area, one should go to the indigenous people, the native Amazonians. Indeed long before the discovery of the Amazon by Europe, the Indians of the várzea including those of Marajó Island at the mouth of the great river, had built up large settlements, practised intensive agriculture, and developed sophisticated cultures (Roosevelt 1991). Very few indigenous várzea peoples, however, survive in identifiable tribal societies. The present inhabitants of várzea lands and the inheritors and transformers of indigenous traditions are the present-day *ribeirinhos* (or *caboclos*) of Brazil and the *ribereños* of the Peruvian Amazon.

The ribeirinhos and ribereños are largely smallholders whose culture is a rich and intricate blend of Amazonian, European, and African traditions. After generations in the várzea, ribeirinhos have developed management patterns and strategies that allow them to make a living in a difficult environment while maintaining the productivity and resilience of várzea ecosystems.

A Diversity of Patterns

The várzea stretches for about three thousand miles from the headwaters to the estuary. The resource-use patterns employed by ribeirinhos/ribereños along this length are many and it is difficult to generalize adequately about them. Perhaps the most important common characteristics of those local practices are their diversity, complexity, and dynamism. Most ribeirinho smallholders employ an array of resource-management techniques that commonly include several types of farming, agroforestry, forest extraction, hunting, fishing, and occasional wage labour. Many of the strategies and techniques they employ are multi-staged and complex. And despite a persistent outside view of rural folk as tradition-bound and slow to change, várzea farmers have continually shown that they are eager for change and capable of responding quickly and appropriately to both new problems and opportunities whether environmental, economic, or political in nature.

To some degree the diversity, complexity, and dynamism of resource management is attributable to the environments that várzea dwellers manage. Often discussed as if it were one kind of environment, we now have come to appreciate that várzea areas differ greatly. For instance, near the mouth of the Amazon, tides flood vast expanses of várzea twice daily, but seasonal variations in flood height are not large. This flood regime presents both different opportunities and problems to farmers than do upper Amazon sites where annual floods commonly raise water levels by 10 metres or more (Padoch and Pinedo-Vasquez 1991; Denevan 1984).

Great geomorphic and ecological diversity can also be found within a very small area of the várzea. Figure 1 presents an idealized (but not unusual) cross-section of a várzea area in the upper part of the Amazon basin. You can clearly distinguish four agriculturally important land-forms in this small section. Three of these land-forms are located within the floodplain: *barreal*, or mudflat; *restinga*, or natural levee; and *aguajal*, or backswamp dominated by the palm *Mauritia flexuosa*. The *altura* (*terra firme*), or area above the floodplain, is here classified as only one zone, although it does include ecologically different sites and micro-gradients (cf. Denevan 1984). While differences in elevation between these sites may be small, frequently the differences in their potential for agricultural production and the risks they present are very great.

Diversity, Complexity, and Change: The Case of Santa Rosa

While figure 1 is indeed an idealized cross-section, it resembles quite closely the várzea near Santa Rosa, a community of about 350 ribereños situated along the edge of the várzea on the Ucayali River, a large tributary of the Amazon in Peru. Santa Rosa seems to be an unexceptional village and indeed it resembles innumerable other communities scattered along the rivers of the Peruvian Amazon.[1] Each one of its residents would identify him or herself as a ribereño/a; they all speak Spanish, all engage in a combination of agricultural, extractive, and other activities, all are descended from families that have lived along Amazon rivers for many generations, and all are poor. There are, however, very significant differences among the villagers. Residents of Santa Rosa trace their heritage to at least five different ethnic groups (Padoch and de Jong 1990), and they differ greatly in their life experiences. Some have spent much of their lives as urban dwellers and travellers, others have stayed very close to home. Households also vary greatly in just how long they have been part of this particular Amazonian community.

This great diversity hidden under a seeming uniformity also applies to the agricultural and agroforestry techniques used by the people of Santa Rosa. The educated, urban folk of this and many other areas of the tropics would

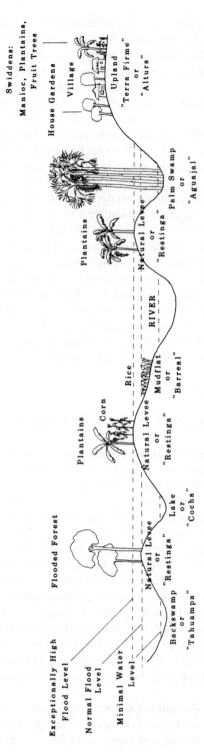

Figure 1 Idealized Cross-section of Floodplain, or "Várzea," of the Ucayali River (Not to scale)

probably dismiss the Santa Rosinos as primitive slash-and-burn farmers; the reality is dramatically different. Exploiting the four different land-forms mentioned above: barreal, restinga, aguajal, and altura, as well as another, the *playa*, or rough-textured sand bar, which is not pictured on figure 1, the farmers of Santa Rosa have developed 12 distinct ways of farming. These range from the highly risky but very productive farming of rice on barreales where production of 4 metric tons per hectare with no fertilization and minimal labour is not uncommon, to highly diverse, long-term and also profitable swidden-fallow agroforestry on the altura. Seemingly inhospitable environments like the playa are used for short-term plantings of cowpeas (*Vigna spp.*) and fertile ones like restingas are used for many production types, including monocultures of corn and manioc, as well as large fruit and even timber plantings.

Not only have Santa Rosinos developed numerous kinds of agriculture, they combine these types into many different patterns. Several years ago we found that the 12 kinds of agriculture available were combined by the 46 individual families in 39 distinct ways. Research showed that there is no one type of agriculture that is used by every household in Santa Rosa and that there is no universally employed strategy. Santa Rosinos do not even agree on whether it is better to diversify resource use, or to specialize in one or two types. Approximately half of Santa Rosa households used four different production types in 1985. A small but significant number – three – however, chose to employ only one production type.

Factors determining which strategy was followed by any household were many and complex. Generally we found that differing access to some kinds of land was important, as were differences in amount and types and the seasonality of labour available to the family. Other important variables were contrasts in access to capital and some capital resources, credit, and opportunities to work beyond the village. Also significant was the family's recent farming history, including losses of crops to floods, as well as general preferences and attitudes. Some farmers were more interested in new crops, techniques, and products, others were less interested in these matters and could be considered more averse to taking risks.

No environment remains static and thus an ability to change is important for any farmer. The Ucayali várzea, however, is characterised by such dramatic seasonal changes that exploiting this environment demands extraordinary flexibility. Any annual flood may change important agricultural land-forms in size, elevation, soil texture, presence of weeds, and spatial relationship with the river. Changes in social factors are no less important. Credit availability, market prices, wage-labour opportunities, and transport facilities change frequently. Families also change in size, health, and in their needs, preferences, and obligations. Any of these factors may affect a household in its decisions on what to plant and where, what fields to continue to manage, and how to alter their management.

Approximately a year after we first studied the households of Santa Rosa, we resurveyed 12 households to see how their strategies had changed. We found then that only three of the families sampled followed the same strategy they had followed the previous year. Nine, or 75 per cent, of the households had adopted a new strategy.

As illustrated by the behaviour of farmers of Santa Rosa, várzea agriculture is not simple, nor uniform, nor unchanging. Another example from the Peruvian Amazon will serve to show how ribereño traditions of forest management are also neither uncomplicated nor static.

Managing Várzea Forests

Near the confluence of two of Peru's greatest rivers, the Amazon and the Napo, is a large area of várzea where Peruvian ecologist Miguel Pinedo-Vasquez investigated previously unstudied ribereño techniques of managing important plywood species in várzea forests. The villagers of Marupá and neighbouring communities of this várzea are similar in many ways to the farmers of Santa Rosa. They specialize, however, in the management of several types of várzea forests along the Napo–Amazon floodplain. *Capinurales*, that is forests dominated by the tree known locally as "capinuri" (*Macruira coriaceae*), are some of the most economically and ecologically important forests in this area. Residents of the Napo–Amazon floodplain extract several products from capinurales for both domestic use and occasional sale in regional markets. The most important is plywood from the capinuri; other products include: resins, edible fruits, medicinal substances from several species, as well as a variety of other construction materials and firewood.

Capinural forests mainly occupy low natural levees along the Napo–Amazon várzea floodplain. Much of this floodplain is periodically farmed, and seedlings of capinuri tend to become established quickly in farm fields, as are juveniles in fallows. Although seedlings, juveniles, and adults respond well to natural and human disturbances, dense and economically important stands of capinuri are established after a long process of management and the application of a complex of management techniques.

The management of capinurales takes place at many stages of forest development: in agricultural fields, in young fallows, and in mature capinural forests. At each stage several management techniques are employed. Each of these techniques is composed of multiple, distinct operations.

The entire management process is too complex to be described here;[2] I will outline very briefly the main operation of the last 10 of the three stages. At this final stage, when capinurales have become mature forests, two techniques are most commonly used to increase their value: *anillado* and *desangrado*. Anillado involves the killing of selected stems of competitor species using girdling of undesirable trees and fire. The technique comprises five activities: selection, marking, removal of understorey vegetation (principally vines), girdling, and burning. The last requires burning a small area of the trunk from which bark was previously removed until the sapwood is affected. The anillado technique usually causes the deselected trees to die rapidly and avoids resprouting from the roots or stem.

Employing the desangrado technique, people remove competitor species, stranglers, and woody climbing vines that may harm valuable trees. The desangrado technique consists of two distinct operations. The first is selection, in which all vines and other plants that are climbing or strangling the trunk or shading valuable trees are selected for removal. The second, girdling, involves the removal of bark, cambium, and sapwood in a ring extending around the woody vine or tree. From the ring fissure, sap, resins, and water are lost. The abundance of resin or sap attracts ants, termites, and other insects that not only eat the sap but also damage any new sprouts. Ribereños forest managers speed the process further by smearing sweet substances on the fissure, tempting even more insects. The resulting infestation controls sprouting of vines and helps kill the competitors.

These two management techniques are applied in managed capinurales on an average of once every six to eight years. By killing vines and selected dominant trees people maintain dense stands of valuable capinuri. The traditional management techniques outlined above apparently succeed in raising the economic value of várzea forests to local villagers with a minimum of labour and capital input. Pinedo-Vasquez found that management by Marupá villagers neither increased nor decreased the number of trees per area found in the forests. The application of these techniques did, however, result in a significant increase in the commercial volume per hectare of timber in capinurales. The mean commercial volume for areas managed as mature capinurales for 16 years was 89 m³/ha, an increase of 65 per cent over the mean of 54 m³/ha for unmanaged capinurales. This increase in commercial value was produced with the application of very small amounts of labour by Marupá's ribereño forest managers and virtually no capital inputs. The ideal rotation for the capinurales was just 16 years, far lower than the 50 to 60 year rotations used in most modern commercial management schemes.

The application of traditional management techniques also rather surprisingly resulted in a statistically significant increase in the number of species in capinurales. In contrast to conventional forestry, commercial value was raised but the floral diversity of the forest was not reduced.

Managing Bananas and Diseases in the Estuary

At the other end of the Amazon's várzea, close to where the Amazon flows out to the sea, lies the village of Ipixuna. The residents of this estuarine region have less-direct ties both genetically and culturally to indigenous Amazonians than do most ribereños of Peru. Some of the present residents are descended from Africans who came into the area as slaves in the early nineteenth century, others are more recent immigrants from Brazil's arid north-east. They share, however, with their Peruvian counterparts a rich tradition of sophisticated resource management of várzea resources. Like the ribereños they actively experiment with applying traditional techniques to managing new problems.

Ipixuna and neighbouring communities were until recently major exporters of bananas. Not only did these smallholders supply the urban markets of the state of Amapá with bananas, they also exported to the major Amazonian city of Belém. In the last several years, however, banana production in the region has been almost completely wiped out by Moko disease, called locally *febre de banana*. The disease, which is common in many banana-producing areas, can be controlled by a concerted and very expensive campaign of destruction of infected plants, repeated disinfection of all tools, and constant inspection of all plantings (Stover and Simmonds 1987). These control measures are not economically feasible for Ipixuna banana producers, and they involve the use of toxic chemicals. Ribeirinhos have instead observed and experimented locally and in the last few years developed an agroforestry system by which they manage the disease although they do not eliminate it.

Ipixuna farmers who are most successful in maintaining adequate production of bananas are those who do the least weeding. In order to produce bananas in an infested area farmers permit weeds and early successional species to fill the spaces in between banana plants while continuing to remove vines and other strangler species. The ribeirinhos who use this pattern, termed the *emcapoeirado* system, observed that weeding facilitates the spread of Moko disease. Farmers who have significantly reduced the amount of weeding they do in fields that include bananas now produce relatively good crops of banana despite disease attacks; others have lost all or nearly all production. The agroforesters also get an economic return from some of the other species incorporated into their system.

The banana emcapoeirado agroforestry system is but one of many resource-management patterns a team of scientists, technicians, and farmers is now studying in the estuarine várzea in Brazil. Ipixuna is one of the research and training sites of the Amazon Cluster of the PLEC project (Brookfield and Padoch 1994). As part of this international effort my colleagues and I are not only detecting important agricultural systems and monitoring local farmers' fields; we are also designing and implementing relevant experiments and extending knowledge of them to other farmers. In Ipixuna, for instance, Brazilian technicians working with us are evaluating emcapoeirado banana agroforests and contrasting them with others where weeding is more intensive. We are also conducting further experimental studies to understand how the emcapoeirado system of banana production reduces disease infestation and to evaluate its economic viability and its ecological effects.

The banana emcapoeirado agroforestry system is a new adaptation of traditional Amazonian agroforestry and forest management practices.[3] Since these systems are based on long traditional patterns, you might ask why these systems are only now being investigated. These systems seem to have been invisible to many of us largely because their developers and practitioners, smallholders like the ribeirinhos, have rarely been credited with sophisticated and valuable knowledge about production and resource management. The systems have perhaps also been invisible because they are confusing to us, they employ principles that many of our advanced, modern systems have discarded. Among these are toleration and even encouragement of biodiversity, principles that many scientists and development experts have only recently come to value.

I have presented only a minute fraction of the many traditional resource-management systems that exist in the Amazon várzea. The many scientists of the PLEC Amazon research cluster hope in the next few years to be able to tell you of many more and to be able to show how some of the principles embodied in these systems may be transferred to our modern resource-management schemes, managing them both more sustainably and more productively.

Notes

1. This example is described in greater detail in Padoch and de Jong (1992). For descriptions of diversity of production in other villages in the lowland Peruvian Amazon, see works by Hiraoka (1985a,b) and Chibnik (1995).
2. For a detailed discussion of this and other forest-management systems in the Napo–Amazon várzea, see Pinedo-Vasquez (1995) and Pinedo-Vasquez and Padoch (n.d.).
3. Many discussions of Amazonian agroforestry systems, including how Amazonians adapt traditional patterns to modern needs and opportunities, are available. Among these are: Denevan and Padoch (1988); Dufour (1990); Hecht (1982); Irvine (1985); Padoch and de Jong (1987, 1989, 1995); Padoch et al. (1985); Posey (1984).

REFERENCES

Brookfield, H. and C. Padoch. 1994. Appreciating agrodiversity: A look at the dynamism and diversity of indigenous farming practices. *Environment* 36(5):6–11, 37–43.

Chibnik, M.S. 1995. *Risky Rivers: The Economics and Politics of Floodplain Farming in Amazonia*. University of Arizona Press, Tucson.

Denevan, W.M. 1984. Ecological heterogeneity and horizontal zonation of agriculture in the Amazon floodplain. In: M. Schmink and C.H. Woods (eds.), *Frontier Expansion in Amazonia*. University of Florida Press, Gainesville.

Denevan, W.M. and C. Padoch. 1988. *Swidden-fallow Agroforestry in the Peruvian Amazon*. Advances in Economic Botany 5. New York Botanical Garden, New York.

Dufour, D.L. 1990. Use of tropical rainforests by native Amazonians. *Bioscience* 40(9):652–59.

Hecht, S.B. 1982. Agroforestry in the Amazon basin: Practice, theory and limits of a promising land use. In: S.B. Hecht (ed.), *Amazonia: Agriculture and Land Use Research*, 33–71. CIAT. Cali.

Hiraoka, M. 1985a. Mestizo subsistence in riparian Amazonia. *National Geographic Research* 1:236–246.

Hiraoka, M. 1985b. Changing floodplain livelihood patterns in the Peruvian Amazon. *Tsukuba Studies in Human Geography* 3:243–275.

Irvine, D. 1985. *Succession Management and Resource Distribution in an Amazonian Rainforest.* Paper presented at the meeting of the American Anthropological Association, Washington, DC.

Padoch, C., J. Chota Inuma, J. de Jong and J. Unruh. 1985. Amazonian agroforestry: A market-orientated system in Peru. *Agroforestry Systems.*

Padoch, C. and W. de Jong. 1987. Traditional agroforestry practices of native and ribereño farmers in the lowland Peruvian Amazon. In: H.L. Gholz (ed.), *Agroforestry: Realities, Possibilities and Potentials.* Martinus Nijhoff/Dr W. Junk Publishers, Dordrecht.

Padoch, C. and W. de Jong. 1989. Production and profit in agroforestry: An example from the Peruvian Amazon. In: J. Browder (ed.), *Fragile Lands of Latin America.* Westview Press, Boulder.

Padoch, C. and W. de Jong. 1990. Santa Rosa: The impact of the minor forest products trade on an Amazonian place and population. In: G.T. Prance and M.J. Balick (eds.), *New Directions in the Study of Plants and People.* Advances in Economic Botany 8:151–8. New York Botanical Garden, New York.

Padoch, C. and W. de Jong. 1992. Diversity, variation, and changes in ribereño agriculture. In: K. Redford and C. Padoch (eds.), *Conservation of Neotropical Forests: Working from Traditional Resource Use.* Columbia University Press, New York.

Padoch, C. and W. de Jong. 1995. Subsistence- and market-oriented agroforestry in the Peruvian Amazon. In: T. Nishizawa and J.I. Uitto (eds.), *The Fragile Tropics of Latin America: Sustainable Management of Changing Environments,* 226–237. United Nations University Press, Tokyo.

Padoch, C. and M. Pinedo-Vasquez. 1991. Floodtime on the Ucayali. *Natural History* (May):48–57.

Pinedo-Vasquez, M. 1995. *Human Impact on Várzea Ecosystems in the Napo-Amazon, Peru.* Unpublished doctoral dissertation. Yale School of Forestry and Environmental Studies, New Haven.

Posey, D.A. 1984. A preliminary report on diversified management of tropical forest by the Kayapó Indians of the Brazilian Amazon. In: G.T. Prance and J. Kallunki (eds.), *Ethnobotany in the Neotropics.* Advances in Economic Botany 1:112–26. New York Botanical Garden, New York.

Roosevelt, A.C. 1991. *Moundbuilders of the Amazon: Geophysical Archaeology on Marajo Island, Brazil.* Academic Press, San Diego.

Stover, R.H. and N.W. Simmonds. 1987. *Bananas.* Longman Scientific and Technical. Harlow, Essex, UK.

8

Global Environment and Population Carrying Capacity

Shunji Murai

Introduction

Uncontrolled population increase appears as the biggest crime of human-kind, because it induces both environmental and ecological destruction through excessive deforestation, urbanization, agricultural development, overgrazing, etc.

The development in the twentieth century, particularly after the Second World War, has accelerated towards a crisis of famine and even extinction of the human civilization, as a result of extreme consumption of natural resources.

Although sustainable development has become a common target since the Earth Summit held in Rio de Janeiro in 1992, this goal cannot be achieved without very severe controls based on a mutual understanding supported by a new philosophy. It will be impossible to continue the current prosperous, but aggressive, development that has been experienced in the twentieth century.

Criteria for Sustainable Development

Dennis L. Meadows presented his paper entitled "It Is Too Late to Achieve Sustainable Development, Now Let Us Strive for Survivable Development" at the Eighth Toyota Conference held in Mikkabi, Japan, in November 1994 (Meadows 1995). According to his scenario, the current situation would lead to "overshoot and collapse," based on the present trends in population increase, industrial production, metal consumption, and grain production.

Table 1 Criteria for Sustainable Development

Human activity	Sustainable	Critical	Destructive
Population increase	<0.5% p.a.	1.0–1.5% p.a.	>2% p.a.
Economic development	3% < GNP < 5%	8% < GNP < 10%	GNP > 10% (overdevelopment) GNP < 0% (underdevelopment)
Deforestation rate	<0.1% p.a.	0.5–1.0% p.a.	>1% p.a.
Forest coverage	>30%	15–20%	<10%
Agricultural development	>0.3% ha per capita	0.15–0.2 ha per capita	0.1 ha per capita
Self-support ratio	>90%	60–70%	<50%
Urbanisation			
Population density	<50 per ha	100–150 per ha	>200 per ha
Population of a city	<0.5 million	>1 million	>10 million

I fully agree with this prediction despite its "non-scientific" approach. The author makes an attempt to define the three criteria of "sustainable," "critical," and "destructive" development from his personal experience, even though the view is not yet scientifically verified. Table 1 summarizes the criteria for sustainable development.

Population Increase

If one reviews the rate of agricultural development or grain production, population increase of over two per cent per year would be destructive. The rate of less than 0.5 per cent may be sustainable or supportable with a healthy development. A rate of 1.0–1.5 per cent will be critical for sustainability.

Economic Development

Overdevelopment of a GNP increase of over 10 per cent per year, as well as underdevelopment with a negative growth rate, will have a destructive impact on the consumption of natural resources and pollution. A GNP increase of 3–5 per cent per annum would be ideal for stabilizing the economy.

Deforestation Rate

An annual deforestation rate of more than one per cent has already been shown to be destructive by the examples of the Philippines in the 1970s and Thailand in the 1980s. On the other hand, for example in the nineteenth-century Malaysia during the colonial period, the deforestation rate was highly controlled and amounted only to less than 0.1 per cent, which was very sustainable. A rate of 0.5–1.0 per cent may be a critical threshold for its sustainability. In my opinion, each country should aim to retain more than 30 per cent of forest cover for stabilizing the climate, groundwater, anti-disaster capability, soil conservation, agricultural production, etc.

In my view, a country with less than 10 per cent forest land is hopeless. A range of 15–20 per cent forest land is a warning signal.

Agricultural Development

The Japanese experience of complete self-reliance in the Edo period demonstrated the sustainability of more than 0.3 hectares per capita given for agriculture, even though the country was not very rich. With less than 0.1 hectare per capita it would be very difficult to maintain even the minimal nutrition level. Note that the world average has decreased from 0.25 hectares per capita in the 1950s to 0.15 hectares per capita in the 1990s, which may be critical for a sustainable level in the future.

Self-support Ratio

Japan has only a 33 per cent self-support ratio, which results in the import of 27 million tons of grain, or 77 per cent of its grain. South Korea imports 68 per cent of its grain, while the figure for Taiwan is 74 per cent. Now China is starting to import grain. Lester R. Brown and Hal Kane estimate in their book *Full House* that in the year 2030 there will be a large gap between grain production and consumption; 204 million tons only in the four largest countries, the USA, China, India, and the former Soviet Union (Brown and Kane 1994).

A self-support ratio of over 90 per cent will be sustainable, while less than 50 per cent will be hopeless at the time when grain production becomes insufficient on a global scale.

Urbanisation

Immigration into big cities has been a critical transition in the twentieth century, because overdevelopment and rapid industrialization cause high consumption of energy and resultant pollution. The urban environment can be maintained very well with a population density of 50 persons per hectare, while over 200 persons per hectare will produce a concrete jungle. The population of a city should be less than 0.5 million for a comfortable environment, while over 10 million in a city is against humanity, with severe pollution and inconvenience.

Scenario of Sustainable Development

There will be two scenarios for carrying capacity presented below.

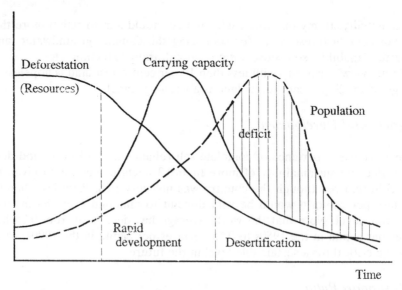

Figure 1 Uncontrolled Development, "Overshoot and Collapse"

Scenario 1: Uncontrolled Development

As shown in figure 1, consumption of natural resources, including deforestation, is very rapid. In the present in developing countries it pushes up
the carrying capacity very quickly, but only in the short term. In the longer
term, it will have drastic environmental and ecological consequences. The
population will increase as the carrying capacity goes up and will continue
even after it starts to decrease. The result will be a deficit between population and the carrying capacity, which may result in famine or refugees.
Desertification will progress for many long years leading to extinction of
civilization. There is much evidence for this in human history: consider for
example Egypt, Greece, Mesopotamia, etc.

Such a pattern is called "overshoot and collapse." The world is presently
on this route.

Scenario 2: Sustainable Development

As proposed in table 1, it is very difficult to meet the criteria for sustainable
development. In order to do so, consumption of natural resources and
industrial production should be controlled.

The increase of carrying capacity will be much more moderate, which will
allow a gentle curve of population increase, and finally an equilibrium, as
shown in figure 2. The birth and the death rate may be almost the same at an
equilibrium, with an increased number of aged persons. The goal of reaching such a curve is not attractive to ambitious politicians, but we should
establish a consensus among nations for sustainable development.

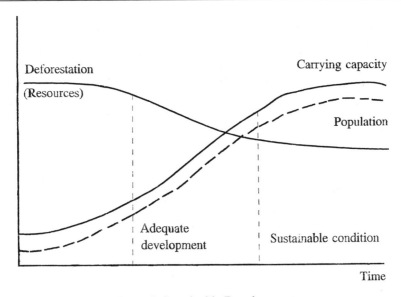

Figure 2 Sustainable Development

Supportable Human Population

The limit for human population is set with the constraints of crop produc-tion. A question arises as to how much population can be potentially sup-ported with respect to crop production. In order to answer the question, the potential crop land must be estimated. According to the author's philoso-phy, new crop land should be developed only in areas of existing grassland, but not in forested areas, because further deforestation will cause a drastic decrease of carbon stock, further accelerating the greenhouse effect.

Table 2 shows an estimate of potential crop land that has been computed by the author and his colleagues (Murai et al. 1990) for five continents, as well as the existing agricultural land and forest area.

The Food and Agriculture Organization of the United Nations (FAO) gives out annually statistics of agricultural production, in particular related to the main crops that are important for the survival of mankind. Based on FAO statistics, the existing and potential agricultural productivity in calories were calculated (table 3).

As seen in table 3, there is potential still to increase agricultural produc-tion by a factor of 1.56. The potential habitability of the earth can be esti-mated with the information on the amount of food energy in kcal that is necessary to support one person. In this study, three different cases of energy consumption, as shown in table 4, were considered in order to esti-mate the potential for supporting a human population on earth.

The current world population, as of 1994, is about 5.4 billion. This implies that we have already passed the earth's carrying capacity if everybody should maintain North American food-consumption levels. If the entire

Table 2 Potential Arable Land, Existing Agricultural Land, and Forest Area (Source: Murai 1995) (1,000 km²)

	Africa	North America	South America	Asia Europe	Oceania	World
Potential arable land	4,681	4,604	1,446	8,832	1,812	21,375
Existing farm land	1,688	2,671	1,163	7,752	478	13,732
Forest area	8,383	8,038	12,221	15,221	1,586	45,774

Table 3 Existing and Potential Agricultural Productivity (10^{12} kcal)

	Africa	North America	South America	Asia Europe	Oceania	World
Existing prod. (A)	293	1,369	361	3,643	68	5,734
Potential prod. (B)	857	2,533	702	4,407	424	8,923
B/A	2.93	1.85	1.95	1.21	6.23	1.56

Table 4 Potential Carrying Capacity Based on Three Levels of Consumption (10^9 people)

Case	Food consumption	Population potential
1	At present levels for each country	7.34
2	World average	8.97
3	North American standard	4.01

world population would assume the current consumption levels, the limit to the world population would be only 7.34 billion people. Even at this level, the world population level will be saturated in the coming about 20 years.

Conclusions

A guideline for achieving sustainable development has been proposed by the author, with criteria and a scenario for sustainability.

The global habitability was estimated with the use of the global vegetation index data in terms of agricultural productivity. If only present grassland can be converted to agricultural land, without deforestation, the world can support a total population of 7.34 billion people under the condition that each country retains the present food-consumption levels. Supposing that the current rate of population increase, 1.7 per cent per annum, continues, the world's population carrying capacity will become critical in only about 20 years.

REFERENCES

Brown, L.R. and H. Kane. 1994. *Full House: Reassessing the Earth's Population Carrying Capacity.* W.W. Norton, New York.

FAO. 1987. *Production Yearbook.* Food and Agriculture Organization of the United Nations, Rome.

Meadows, D.L. 1995. It is too late to achieve sustainable development, now let us strive for survivable development. In: S. Murai (ed.), *Toward Global Planning of Sustainable Use of the Earth: Development of Global Eco-engineering,* 359–374. Elsevier, Amsterdam.

Murai, S. 1995. Development of Global Eco-engineering using Remote Sensing and Geo-information Systems. In: S. Murai (ed.), *Toward Global Planning of Sustainable Use of the Earth: Development of Global Eco-engineering,* 1–11. Elsevier, Amsterdam.

Murai, S., Y. Honda, K. Asakura and S. Goto. 1990. *An Analysis of Global Environment by Satellite Remote Sensing: What Population Can the Earth Feed?* Institute of Industrial Science, University of Tokyo.

9

Concluding Remarks

Juha I. Uitto

Today's world is experiencing a growth in the human population that is unparalleled in history. The world's population has doubled since the 1950s and we have now reached a level of more than 5.6 billion people on this planet. Every year, some 90 million new people are added to this number. The "most likely" scenario forecast by the United Nations puts world population at 10 billion in the year 2050.

The pressures placed on agricultural production on a global scale are daunting. In 1961, the global arable area was 1.3 billion hectares, of which 10 per cent was irrigated. Thirty years later, the global arable area has increased by only 100 million hectares, while the share under irrigation has gone up to 17 per cent. However, even if the total arable area has increased somewhat, if one adjusts it by the number of people that have to be fed, the figure looks very different. Since 1961, the amount of arable land per capita has decreased from 0.44 hectares to a mere 0.27 hectares in 1990 (Engelman and LeRoy 1995). This already small amount of a quarter of a hectare per head is rapidly shrinking.

It is, thus, obvious that the mere increase in global population is putting fierce pressures on the ability of the Earth to feed its inhabitants. In the past decades, food production has been able to cope with the increase in the number of consumers through constantly higher output per area unit. The so-called green revolution has been instrumental in increasing the crop yields manifold and, thus, warding off hunger in many countries. However, there is now clear evidence that the yield increases have reached or will soon reach their limits, with most farmers already using improved varieties and techniques.

Take rice, for example, on which most Asians, and a large share of others

84

in different parts of the world, rely for their staple food. Rice farming in Japan is very intensive and has served as a model to emulate for many nations. Japan's rice yields per area unit were constantly rising for the past century. This rise, however, came to a halt in 1984 after which the yields have actually fallen slightly. Evidence from other rice-growing countries, including Bangladesh, India, the Philippines, and Thailand, suggests that intensification of land use and continuous cropping of rice with two or three yields per year has actually led to a decline in unit yields (Brown and Kane 1994).

Consequently, further increases in yields will be difficult to attain without significant new technological advances. Furthermore, the "green revolution" is associated with other problems, as well, including the excessive need for irrigation water, and the reliance on fertilizers, pesticides, and other agricultural chemicals with adverse environmental effects. Similarly, usage of modern high-yielding crop varieties leads inevitably to the loss of the biological diversity inherent in traditional agricultural systems, rendering the farms more vulnerable to pests and climatic variation, at the same time as reducing the gene pool available to mankind.

With little new arable land available for clearing, farmers will mostly have to rely on existing land or, as is frequently the case, move to increasingly marginal areas. The latter increases the vulnerability of the farm land to erosion and is often associated with loss of topsoil and the encroachment of deserts.

In South-East Asia, earlier UNU research identified the inevitable population growth and the rapid economic growth as the basic constraints within which any improved environmental management must evolve (Brookfield and Byron 1993). These basic constraints form the main driving forces of environmental change in the region. The research further identified a number of present trends that are clearly unsustainable in the long run. These include the present methods of intensification of agricultural production, which are heavily dependent on the extensive use of agricultural chemicals; most land-use practices in upland areas, the importance of which will only increase in the future; urbanization with the associated lack of adequate pollution control and waste management; the increasing use of energy; and forestry practices that do not replace depleted resources, leading to the loss of biological diversity at an alarming rate.

However, even if on global and regional scales the trends are clear, at the local level there are considerable variations in the interlinkages between population growth, agricultural production, and the sustainability of the environment. Whereas in some places environmental degradation caused by over-cultivation and over-grazing has led to the destruction of the agricultural production base, causing entire villages to become environmental refugees, in others increases in smallholder population have actually led to more sustainable production practices. Such cases have recently been reported for example from Kenya in East Africa, where people reacted to the increasing population pressure by adopting improved farming practices, building terraces to fight erosion and planting trees for fuelwood (Tiffen et al. 1994).

This example demonstrates the complexity of the issue. The simplistic and deterministic notions that so often figure in the discussions and even are used as bases for policy formulation need to be modified to reflect the complex and varying conditions encountered in reality.

The Fourth UNU Global Environmental Forum was largely based on the detailed research work carried out within the international collaborative research programme on "People, Land Management, and Environmental Change," or PLEC as it is called in short, undertaken under the auspices of the United Nations University. PLEC is aimed at a systematic field-level analysis of sustainable land management and agrotechnology, and the maintenance of biological diversity in small-farm regions in the tropical and subtropical parts of the world. A basic premise of the programme is that the indigenous farming practices in the various parts of the world are frequently well-adapted to the prevailing ecological conditions and contain solutions to the issue of sustainability. The traditional farming systems are, however, under severe pressure due not only to population growth, but to various societal forces. Detailed studies are needed to understand the process and to provide policy-relevant advice to the concerned countries and organizations involved in agricultural development.

PLEC lays particular stress on the diversity of farming practices over small areas, their responsiveness to changing conditions, and adaptation to ecological variability. Cases covered at this Forum highlighted the widely varying environmental, social, economic, and political conditions that pose the framework for sustainable agricultural development in such different places as northern Thailand, Papua New Guinea, and the Amazonian floodplains. In addition, PLEC operates in other regions of the world not represented here today, including the Caribbean islands; the dynamic West African forest–savanna transition zone; a selection of sites in the three Eastern African countries of Kenya, Tanzania, and Uganda; and the Chinese province of Yunnan.

In addition to the geographical distribution, importance is placed on the multidisciplinary approach. The Forum speakers come from backgrounds ranging from agriculture and the soil sciences, through geography to anthropology. Knowing the ground truth is essential, but modern tools such as satellite remote sensing and Geographical Information Systems (GIS) can be utilized to provide information on a higher scale. The combination of these approaches is indispensable to the fuller understanding of the conditions.

Other UNU research complements the PLEC programme. Notably, the long-running research programme on "Mountain Ecology and Sustainable Development" has related objectives focusing on the man–land interaction in the mountainous and highland areas of the world. Policy-oriented work is being carried out in the Hengduan mountains of China, in the Himalayas, in the mountains and highlands of Africa, and in the Andes in Latin America (e.g. Ives and Messerli 1989; Ives and Uitto 1994).

The results of UNU research are being actively fed into the policy-making process by the United Nations. The University's work is also geared towards greater capacity building in developing countries to allow the countries to better formulate and implement their policies concerning sustainable development.

The quest for sustainable development is an awesome task and it requires the participation of all sectors of society. We hope that our cooperation within research and capacity building will make a modest contribution to this task.

REFERENCES

Brookfield, H. and Y. Byron (eds.). 1993. *South-East Asia's Environmental Future: The Search for Sustainability*. United Nations University Press, Tokyo, and Oxford University Press, Kuala Lumpur.

Brown, L.R. and H. Kane. 1994. *Full House: Reassessing the Earth's Population Carrying Capacity*. W.W. Norton & Company, New York.

Engelman, R. and P. LeRoy. 1995. *Conserving Land: Population and Sustainable Food Production*. Population and Environment Programme, Population Action International, Washington, DC.

Ives, J.D. and B. Messerli. 1989. *The Himalayan Dilemma: Reconciling Development and Conservation*. The United Nations University, Tokyo, and Routledge, London.

Ives, J.D. and J.I. Uitto. 1994. Mountain ecology and sustainable development. *Global Environmental Change: Human and Policy Dimensions* 4(3):261–64.

Tiffen, M., M. Mortimore and F. Gichuki. 1994. *More People, Less Erosion: Environmental Recovery in Kenya*. John Wiley & Sons, Chichester, UK.

Contributors

Dr. Harold Brookfield
Department of Anthropology
Division of Society and Environment
Research School of Pacific and Asian
    Studies (RSPAS)
The Australian National University
Canberra, Australia

Dr. Richard A. Meganck
Director
United Nations Environment
    Programme
International Environmental Technol-
    ogy Centre (UNEP/IETC)
Osaka, Japan

Dr. Janet Henshall Momsen
Department of Applied Behavioral
    Sciences
University of California
Davis, USA

Dr. Shunji Murai
Institute of Industrial Science
University of Tokyo
Tokyo, Japan

Dr. Ryutaro Ohtsuka
Department of Human Ecology

School of International Health
Faculty of Medicine
University of Tokyo
Tokyo, Japan

Ms. Akiko Ono
Japan Wildlife Research Center
Tokyo, Japan

Dr. Christine Padoch
Curator
Institute of Economic Botany
The New York Botanical Garden
New York, USA

Dr. Kanok Rerkasem
Research Scientist
Multiple Cropping Center
Faculty of Agriculture
Chiang Mai University
Chiang Mai, Thailand

Dr. Graham Sem
Lecturer
Department of Geography
The University of Papua New Guinea
Port Moresby, Papua New Guinea

Dr. Michael Stocking
School of Development Studies
University of East Anglia
Norwich, United Kingdom

Dr. Juha I. Uitto
Academic Officer
The United Nations University
Tokyo, Japan